Voodoo Science

Voodoo SCIENCE

The Road from Foolishness to Fraud

ROBERT L. PARK

OXFORD
UNIVERSITY PRESS

2000

OXFORD
UNIVERSITY PRESS

Oxford   New York
Athens   Auckland   Bangkok
Bogotá   Buenos Aires   Calcutta
Cape   Town   Chennai   Dar   es
Salaam   Delhi   Florence   Hong
Kong   Istanbul   Karachi   Kuala
Lumpur   Madrid   Melbourne
Mexico   City   Mumbai   Nairobi
Paris   São   Paulo   Singapore
Taipei   Tokyo   Toronto
Warsaw

and associated companies in
Berlin   Ibadan

Published by Oxford University
Press, Inc.
198 Madison Avenue, New York,
New York 10016

Oxford is a registered trademark
of Oxford University Press

Library of Congress Cataloging-in-
Publication Data
Park, Robert L.
Voodoo science : the road from
foolishness to fraud / by Robert L.
Park
p.   cm.
Includes index.
ISBN 0–19–513515–6
Science—Social aspects—United
States. 2. Fraud in science—
United States. I. Title
Q175.52.U5P37   2000
509'.73—dc21   99-40911

Book   design   by   Adam   B.
Bohannon

9 8 7 6 5 4

Printed in the United States of
America
on acid-free paper

CONTENTS

PREFACE

IN 1982, WILLIAM (WILLY) FOWLER, a Cal Tech physicist whose seminal work on elemental abundances would be recognized with a Nobel Prize a year later, called me to ask if I would use my sabbatical year to establish an office of public affairs in Washington for the American Physical Society. Physicists needed to be kept informed of developments in Washington that were having a profound effect on them and the things they value. Perhaps, he said, it would also be possible to communicate the concerns of the physics community, not just to the leaders of government but to the public.

It was to be an experiment. Through most of its existence, the American Physical Society, then headquartered in New York, had not felt the need for a Washington presence, but times were changing. Public support for science had begun unraveling during the Vietnam War. Scientists who had enjoyed public adulation for their contributions to victory in World War II and for putting a man on the Moon found themselves denounced for government connections that a short time earlier were seen as patriotic. Within government, the Cold War ruled. Government censors were attempting to control the exchange of unclassified scientific papers at open scientific meetings, the federal budget for fundamental research had been slashed, and the nuclear arms race was spiraling out of control.

I was long overdue for a sabbatical. There had just never been a time when I felt I could be away: graduate students needed guidance, there were the constant demands of proposal writing to keep a large research group going, I edited an international journal in the field of surface physics and chaired the Department of Physics and Astronomy at the University of Maryland. In the spring, however, Ellen Williams, who had joined my research group from Cal Tech as a post-doc, was appointed an assistant professor. I could leave the group in her hands for a year. At the end of the year someone else would take over in Washington; I would return to teaching and to the research on the atomic structure of crystal surfaces that had occupied most of my waking hours for more than twenty years.

At the end of the year, however, there was no one standing by to take my place in Washington. There is no greater privilege than to be a tenured professor of physics at a great university, but tenure carries with it an obligation to speak out against the errors of our times. I elected to divide my time between teaching at the University of Maryland and the direction of the Washington office. My direct involvement in research wound down as my graduate students completed their research. My research group long ago became Ellen's group, prospering under her direction.

Of the major problems confronting society—problems involving the environment, national security, health, and the economy—there are few that can be sensibly addressed without input from science. As I sought to make the case for science, however, I kept bumping up against scientific ideas and claims that are totally, indisputably, extravagantly, wrong, but which nevertheless attract a large following of passionate, and sometimes powerful, proponents. I came to realize that many people choose scientific beliefs the same way they choose to be Methodists, or Democrats, or Chicago Cubs fans. They judge science by how well it agrees with the way they want the world to be.

A best-selling health guru insists that his brand of spiritual healing is firmly grounded in quantum theory; half the population believes Earth is being visited by space aliens who have mastered faster-than-light travel; and educated people wear magnets in their shoes to restore natural energy. Did we set people up for this? In our eagerness to share the excitement of discovery, have scientists conveyed the message that the universe is so strange that anything is possible? What can we tell people that will help them to judge which claims are science and which are voodoo?

I began to include my encounters with voodoo science in my weekly electronic column, *What's New*, and in articles in the popular press. One of those articles was an op-ed in the *New York Times* to which the editor gave the title "Voodoo Science." Both the article and the title seemed to strike a resonance with readers, and I was invited to expand it for inclusion in a wonderful book of essays, *Dumbing Down: Essays on the Strip-Mining of American Culture*, edited by Katherine Washburn and John Thornton at W. W. Norton & Company. My agent, Theresa Park (no relation), urged me to further expand "Voodoo Science" into a book. She has remained a constant source of encouragement and support, even through a disastrous first attempt.

I will, of course, be delighted if scientists read my book and find it entertaining, but it wasn't written for them. I had no interest in writing a scholarly book to be read only by other scholars. Kirk Jensen, my editor at Oxford, agreed, and suggested that the book read, as much as possible, like a narrative, unencumbered by references and footnotes. The price for that approach is that it leaves scant opportunity to acknowledge the writings of others, particularly the clear-eyed champions of a rational, scientific view of the universe, including Richard Dawkins, Martin Gardner, Ursula Goodenough, Steven Gould, James Randi, Michael Shermer, Steven Weinberg, and E. O. Wilson.

I am indebted to those who took time from their own writing to read all or portions of the manuscript: Barry Beyerstein, K. C. Cole, Alex Dessler, Ursula Goodenough, Francis Slakey, David

Voss, and Peter Zimmerman. They suggested numerous improvements and spared me certain embarrassment. I bear sole responsibility for remaining flaws. Special thanks to my son, Robert T. Park, who helped me think through many of the ideas in the book during long Sunday runs along Northwest Branch. Finally, I am grateful to the American Physical Society, which for sixteen years has allowed me to share my thoughts with an audience of some of the smartest people in the world, asking only that my weekly column carry a disclaimer: "Opinions are the author's and are not necessarily shared by the APS, but they should be."

Voodoo Science

ONE
IT'S NOT NEWS,
IT'S ENTERTAINMENT
In Which the Media Covers Voodoo Science

JOE NEWMAN AND THE
ENERGY MACHINE

I CALLED JOE NEWMAN at his home in Lucedale, Mississippi. I was surprised when he answered the phone; I had tried several times before and always got a recorded message offering his book, *The Energy Machine of Joseph W. Newman*, for $74.95. I explained that I was writing a book about ideas that are not generally accepted by scientists, and it would not be complete without a full account of the Energy Machine. He seemed suspicious. "It's all in my book," he snapped. I told him I'd read his book at the time of his 1986 Senate hearing, but I wondered if his ideas had changed over the years. His voice softened. Well, he said, the book had been expanded and I should buy a new copy, but he still stood by everything he'd said before about how the Energy Machine works.

I waited through a long silence while he thought about what else he should tell me. Then he began talking. The big change since the Senate hearing was that Joe Newman had found God. Raised in a Methodist orphans' home until he ran away at fourteen, Joe became an atheist because he didn't believe a God would permit little children to suffer that much. But he now realized that the Energy Machine was meant to relieve human suffering, and that God had chosen Joe Newman to make the discovery because "he knew Joe Newman would be a good steward for his gift."

It saddened Newman that, in spite of his efforts, the benefits from the Energy Machine were still not reaching the people of the world. "I do it for the human race," he said, "but the people I trusted most betrayed me and the human race." His patent lawyer, the company that supplied the batteries for his machine, even those who had testified on his behalf in court and in the Senate hearing had all used or sold his ideas. There are motors based on his ideas on the market right now, he assured me, that are more than 100 percent efficient; the manufacturers refuse to admit it so they won't have to pay him royalties. He didn't mind his ideas being stolen, if that meant they would be turned into things that would help the world, but these people were hiding the truth about his discovery. "They don't care about humanity," he said sadly.

There was a flash of the old Joe Newman when he vowed to sue his betrayers. The refusal of the U.S. Patent and Trademark Office to grant him a patent for "an unlimited source of energy" no longer mattered, he explained; he had patented the Energy Machine in Mexico, and because of the NAFTA and GATT agreements, his patent is now good all over the world. "A jury will bury these people," he assured me.

Meanwhile, a lot of people continue to believe that Joe Newman really has found a way to generate unlimited amounts of energy. Newman told me he was still demonstrating his Energy Machine around Mississippi and Louisiana and appearing from time to time on radio talk shows. He said they like him on talk radio, "not because they believe me, but because I'm good for ratings. Creative people," Joe sighed, "die poor." There was another long pause. "The people of Mississippi have not stood up behind this technology as they should have," he told me. "I'm leaving the state and heading west. People out there are very concerned about pollution; they'll recognize what this technology can do."

How much of this does Joe Newman really believe? Even now it's impossible to tell. Perhaps not even Joe Newman knows. But he has bounced back before. I first heard about him and his Energy Machine on January 11, 1984, on the CBS *Evening News*. "What's the answer to the energy crisis?" Dan Rather was asking, "Suppose a fellow told you the answer was in a machine he has developed? Before you scoff, take a look with Bruce Hall." CBS reporter Bruce Hall had traveled to the rural hamlet of Lucedale. A mile down a dirt road, past the KEEP OUT and NO TRESPASSING signs, Hall stood with Joseph Wesley Newman in front of his garage workshop. He described Newman as "a brilliant self-educated inventor." An intense, handsome man in his forties, dressed in work clothes, his dark hair combed straight back, the plainspoken mechanic looked directly into the eyes of the viewers. He declared that his Energy Machine could produce ten times the electrical energy it took to run it. "Put one in your home," he said, "and you'll never have to pay another electric bill."

It's the sort of story Americans love. A backwoods wizard who never finished high school makes a revolutionary scientific discovery. He is denied the fruits of his genius by a pompous scientific establishment and a patent examiner who rejects his application for a patent on "an unlimited source of energy" without even examining it, on the grounds that all alleged inventions of perpetual motion machines are refused patents. Not a man to be pushed around, Joseph Wesley Newman takes on the U.S. government, filing suit in federal court against the Patent and Trademark Office. It's the little man battling a gigantic, impersonal system.

There was no one on the CBS *Evening News* to challenge Newman's claim. On the contrary, the report included endorsements from two "experts" who had examined Newman's Energy Machine. Roger Hastings, a boyish-looking Ph.D. physicist with the Sperry Corporation, declared, "It's possible his theory could be correct and that this could revolutionize society." Milton Everett, identified as an engineer with the Mississippi Department of Transportation, told viewers, "Joe's an original thinker. He's gone beyond what you can read in textbooks." Watching the CBS broadcast that evening, most viewers must have been left wondering how the Patent Office could be so certain Joe Newman was wrong.

The Patent Office based its judgment on the long and colorful history of failed attempts to build perpetual motion machines, going back at least to the seventeenth century. Waterwheels had been used in Europe for centuries to grind flour, but many areas lacked suitable streams for a mill. Farmers in those areas were forced to transport their grain to distant mills and then lug the meal back. In 1618 a famous London physician named Robert Fludd wondered if a way could be found to run a mill without depending on nature to provide the stream. Dr. Fludd was, like Joe Newman, endowed with boundless self-confidence and a wide-ranging imagination; we will meet him again in his role as a healer. It occurred to Dr. Fludd that the waterwheel could be used to drive a pump, as well as to grind flour. The water that had turned the wheel could then be pumped back up into the millrace. That way, he reasoned, a reservoir of water could be used to run the mill indefinitely.

Dr. Fludd's idea failed, but his failure helped to lead others to one of the greatest scientific insights in history, paving the way for the industrial revolution. The amount of work a waterwheel can perform is measured by the weight of the water that comes out of the millrace, multiplied by the distance the water descends in turning the wheel. For Dr. Fludd's idea to work, the water would have to be lifted back up the same distance it fell in turning the wheel. All the energy generated in turning the wheel would be needed just to raise the water back to the reservoir. There would be no energy left over to grind the flour.

But the concept of energy or "work" as a measurable quantity did not exist in the seventeenth century. It would be another two hundred years before the flaw in Dr. Fludd's machine would be stated in the form of a fundamental law of nature: *energy is conserved*. Written as a mathematical equation, it is known as the *First Law of Thermodynamics*. There is no firmer pillar of modern science. It is the law that explains why a ball, no matter what it's made of, never bounces higher than the point from which it's dropped. The conservation of energy is consistent with our everyday experience: you can't get something for nothing.

Even if it ground no flour, however, Dr. Fludd's waterwheel

will keep running into Joe Newman. Maybe we will come to understand him. But the Energy Machine of Joe Newman is just one small example of the abuse of science. It's all around us.

VOODOO SCIENCE

Science fascinates us by its power to surprise. Unexpected results that appear to violate accepted laws of nature can portend revolutionary advances in human knowledge. In the past century, such scientific discoveries doubled our life span, freed us from the mind-numbing drudgery that had been the lot of ordinary people for all of history, revealed the vastness of the universe, and put all the knowledge of the world at our fingertips. As a new century begins, molecular biology is unraveling the secrets of life itself, and physicists dare to dream of a "final theory" that would make sense of the entire universe.

Alas, many "revolutionary" discoveries turn out to be wrong. Error is a normal part of science, and uncovering flaws in scientific observations or reasoning is the everyday work of scientists. Scientists try to guard against attributing significance to spurious results by repeating measurements and designing control experiments. But even eminent scientists have have had their careers tarnished by misinterpreting unremarkable events in a way that is so compelling that they are thereafter unable to free themselves of the conviction that they have made a great discovery. Moreover, scientists, no less than others, are inclined to see what they expect to see, and an erroneous conclusion by a respected colleague often carries other scientists along on the road to ignominy. This is *pathological science,* in which scientists manage to fool themselves.

If scientists can fool themselves, how much easier is it to craft arguments deliberately intended to befuddle jurists or lawmakers with little or no scientific background? This is *junk science*. It typically consists of tortured theories of what *could be* so, with little supporting evidence to prove that it *is* so.

Sometimes there is no evidence at all. Two hundred years ago, educated people imagined that the greatest contribution of science would be to free the world from superstition and humbug. It has not happened. Ancient beliefs in demons and magic still sweep across the modern landscape, but they are now dressed in the language and symbols of science: a best-selling health guru explains that his brand of spiritual healing is firmly grounded in quantum theory; half the population believes Earth is being visited by space aliens who have mastered faster-than-light travel; and educated people wear magnets in their shoes to draw energy from the Earth. This is *pseudoscience*. Its practitioners may believe it to be science, just as witches and faith healers may truly believe they can call forth supernatural powers.

What may begin as honest error, however, has a way of evolving through almost imperceptible steps from self-delusion to fraud. The line between foolishness and fraud is thin. Because it is not always easy to tell when that line is crossed, I use the term *voodoo science* to cover them all: pathological science, junk science, pseudoscience, and fraudulent science. This book is meant to help the reader to recognize voodoo science and to understand the forces that seem to conspire to keep it alive.

The first exposure of most people to new scientific claims is through the news media, usually television, and that is where our story begins.

THE PATTERSON CELL AND THE MAGIC BEADS

ABC's *Morning News* on February 6, 1996, carried a story about another inventor, James Patterson, and another inexhaustible source of energy. "When Jack planted his magic seeds he got a beanstalk," the ABC news anchor began, "but when inventor James Patterson took his beads—almost perfectly round beads—and mixed them with water, he says he got energy and lots of it. Now his invention may turn out to be the goose that lays golden eggs. Here's Michael Guillen." Once again, a trusted network news show was inviting people to take a "science" story seriously.

Correspondent Michael Guillen was standing with James Patterson in the inventor's cluttered garage workshop. Patterson had achieved a measure of success with a process for making tiny plastic beads that have a variety of rather mundane uses. Dressed in a laboratory smock, the cheerful, white-haired, seventy-five-year-old appeared to be a sort of avuncular caricature of an inventor. "I'm better'n a millionaire. I have converted alchemy to turn little beads into gold," he chuckles. When he coats his polymer beads with nickel and palladium, mixes them with salt water, and passes an electric current through it, he tells Guillen, two hundred times as much energy comes out as he puts in. How does it work? He says he has no idea.

Palladium? Electrolytic cells? I began paying closer attention. Patterson's claim sounded suspiciously like the discredited "cold fusion" claim made seven years earlier by Stanley Pons and Martin Fleischmann, two chemists from the University of Utah. Michael Guillen, who is himself a physicist, must surely have recognized the similarity, but if he did, he didn't share it with his audience. Instead, Guillen put on a serious face and spoke directly to the camera: "There have been dozens of claims of ideal energy sources around the world, but this device is different, because of the inventor's distinguished track record and because it has attracted serious interest from major companies. Most important, it seems to have been confirmed by independent scientists at prestigious universities. But for some, it just doesn't ring true."

That line served as the cue to bring on the "talking heads." These are "experts," usually identified only by a brief caption, delivering short, taped, sound bites. They are a standard fixture in television coverage of science. The first talking head, labeled "John Huizenga, nuclear scientist," said, "I would be willing to bet there's nothing to it." That was it! The contrary point of view in under three seconds. A second gray head, identified as Quentin Bowles, professor at the University of Missouri, disagreed: "It works, but we don't know why it works. That's the bottom line." The entire exchange took only seven seconds.

Quentin Bowles, an engineer, was not well known. John Huizenga, on the other hand, was a distinguished professor of nuclear chemistry at the University of Rochester, member of the National Academy of Sciences, head of a government panel convened in 1989 to investigate the cold fusion claims of Fleischmann and Pons, and author of the most authoritative book on the cold fusion controversy. Guillen could not have found a better-qualified expert. Any scientist who had followed cold fusion knew who John Huizenga was, but most viewers had never heard of either man. This created what Christopher Toumey, in *Conjuring Science,* calls "pseudosymmetry"—the false impression that scientists' opinions are about equally divided on claims that may have little or no scientific support.

I searched the morning papers for any mention of the Patterson cell. Television news is valuable as a heads-up, but it's no substitute for print; with TV you find yourself wondering if a caption said Missouri or Mississippi, but there's no going back to check. In this case, however, I found nothing in the papers. A simple device that produces two hundred times as much energy as it consumes would alter the course of history. You might suppose that James Patterson and his magic beads would be a major news story, yet no other media outlet had mentioned it. And why had Michael Guillen so carefully avoided mentioning cold fusion? Was the public being deliberately misled?

Perhaps. The story wasn't news, and it certainly wasn't science; it was entertainment. Patterson, like Joe Newman, is an appealing and colorful figure. The term *cold fusion* was avoided because it evokes a negative image: most people recall vaguely that cold fusion claims have been discredited. By avoiding the term, however, Guillen obscured the really interesting story: cold fusion is alive! It has not gone away.

Although cold fusion disappeared from the front pages years ago, a dwindling band of believers has gathered each year since 1989 for a meeting at some swank international resort to share the results of their efforts to resuscitate cold fusion. The venue for the 1996 International Cold Fusion Conference was a luxury hotel with its own golf course in Sapporo, Japan. In 1995 the meeting was in Monte Carlo; the year before that, it was Maui. Like previous meetings, the Sapporo conference was a congenial gathering; the conference is not widely advertised outside the tight group of acknowledged believers. Perhaps because they feel themselves besieged by the rest of the scientific community, there is virtually no dissent among them. Even when they seem to be reporting contradictory results, they refrain from open criticism of each other's work and struggle to find common ground. One of the speakers at the Sapporo cold fusion conference was inventor James Patterson.

Those with doubts about cold fusion have not felt comfortable at these meetings and rarely attend. A notable exception is Douglas Morrison, a British-born high-energy physicist from CERN, the great European accelerator laboratory in Switzerland. Morrison has taken it upon himself to attend every one of these meetings and to keep the rest of the scientific community informed about the proceedings. Two years before the Sapporo event Morrison had officially retired, but for him, as for many prominent scientists, that only meant less income and more freedom to pursue whatever he found interesting and important. He seems unperturbed by the fact that he is treated with suspicion and even open hostility by other attendees, many of whom apparently believe he is in the pay of some powerful international organization bent on suppressing cold fusion. Morrison, who pays his own way to these conferences, simply explains that he loves good science—and dislikes bad science. Morrison recalls that Pons, in a 1989 interview, had shown what he said was a small cold fusion boiler: "Simply put," Pons had explained, "in its current state it could provide boiling water for a cup of tea." Each year at the cold fusion conference Morrison politely asks, "Please, may I have a cup of tea?"

The believers return each year hoping for good news, but the news in 1996 was troubling. The principal source of funding for cold fusion research in the United States had been EPRI, the Electric Power Research Institute, which is jointly operated by private power companies. A few months earlier, EPRI had announced that it was terminating support for cold fusion research. Now there were rumors that MITI, the Japanese Ministry of International Trade and Industry, which was sponsoring the Sapporo meeting, had also decided to pull out of cold fusion. To believers, it seemed inexplicable. Why would these organizations withdraw their support just when the research was on the verge of finally unraveling the puzzle of cold fusion?

Alas, it is always on the verge. Each year at the cold fusion conference there is great excitement over new results that are said at last to show incontrovertible proof that fusion is taking place at low temperatures. Perhaps it's new evidence of neutrons or gamma rays characteristic of deuterium fusion; or helium, the product of fusion, has been found in the metal lattice; or at last a reliable experiment has proven there is an energy gain; or a new theoretical analysis has shown that cold fusion is consistent with known laws of physics after all. But by the time of the next meeting, many of these papers will have been discredited or withdrawn because problems were discovered with the equipment, or because a flaw has been found in the theoretical analysis, or because others have been unable to obtain the same result. Cold fusion is no closer to being proven than it was the day it was announced.

These are scientists; they are presumably trained to view new claims with skepticism. What keeps them coming back each year with hope in their breasts? Why does this little band so fervently believe in something the rest of the scientific community rejected as fantasy years earlier? Agent Fox Mulder, the fictional FBI agent in the popular TV drama *The X-Files*, who deals with cases that seem to involve the paranormal, has a poster on the wall of his office that reads simply: I WANT TO BELIEVE. Like agent Mulder, the cold fusion faithful want to believe. If we are to understand why they have selected cold fusion to believe in, we must first review the extraordinary events in the spring of 1989.

COLD FUSION AND THE UTAH CHEMISTS

One reason scientists seemed unable to deal with Joe Newman's claim was that he had followed none of the "rules." New scientific findings are normally shared first with a few close colleagues, and perhaps tested in a departmental seminar. The work may also be formally presented at a scientific conference—although the Superdome might be frowned on as a scientific venue. If no problems turn up, the work is submitted to an appropriate scientific journal for publication. The journal editor will choose a few anonymous experts to review the work for obvious errors in the methods or reasoning and to ensure that proper credit is given to previous work. Reviewing the manuscripts of other scientists carefully and objectively is regarded as a sacred obligation.

Well, that's the theory, anyway. In practice, the process is occasionally noisy and unpleasant. Heated arguments can take place at scientific conferences. Reviewers are sometimes accused of obstructing the publication of results that contradict their own work, and editors are accused of bias. Rivalries develop that are as strong as anything that takes place on the playing field. Foolish work may find its way into print, while a spectacular insight becomes mired in some petty dispute. And yet, overall, the system works amazingly well: good work eventually rises to the top, while the clutter of shoddy science remains manageable. The scientific process transcends the human failings of individual scientists—but with cold fusion, the process was in for a jolt.

It was Thursday, March 23, 1989. The Sun warmed the Earth that day, as it had for five billion years, by the high-temperature fusion of hydrogen nuclei. It will continue doing so for many more billions of years, which is to say that even in the fierce cauldron of the Sun, fusion is a rather slow process, which is a very good thing for us. In Salt Lake City, the University of Utah held a press conference to announce that two chemists, Martin Fleischmann and Stanley Pons, had discovered a limitless, nonpolluting source of energy. They called it cold fusion. "We've established a sustained fusion reaction by means that are considerably simpler than conventional techniques," Professor Pons declared. If it was true, they had duplicated the source of the Sun's energy—in a test tube.

Moshe Gai, a Yale nuclear physicist, remembers he was on the expressway driving home when he heard the news on National Public Radio. If it was true, it would be the scientific discovery of the century. Gai was fast approaching exit 51; there was barely time to maneuver into the exit lane. He crossed over the expressway, reentered going the other direction, and headed back to the university. He thought he knew how to test the Utah claim.

The story had actually broken that morning. Hours before the Salt Lake City press conference, it appeared in the *Financial Times* in London and the *Wall Street Journal*. The *Wall Street Journal* would run unfailingly optimistic cold fusion stories over the coming weeks, and even carry an editorial using the episode to boast that it was the leader in covering new technological developments. The University of Utah was using the *Wall Street Journal* to sell cold fusion, and the *Wall Street Journal* would use cold fusion to sell papers. Such a discovery could spawn an industry larger than any ever seen on Earth; that the story had been leaked to the world's most influential financial dailies was certainly no accident.

But there was an accident that night. At four minutes past midnight, in some cruel prank of the gods, the supertanker Exxon Valdez ran aground on Bligh Reef in Alaska's Prince William Sound, creating the largest oil spill in U.S. history. News of the Exxon Valdez disaster broke too late to make the morning papers on the East Coast, but the discovery of a limitless source of non-polluting energy was on the front page of the *New York Times*. In the days that followed, the tragic images of dying birds and seals coated with thick, black oil would be a daily reminder of the price civilization pays for the energy that propels it. To the defilement of Prince William Sound, add acid rain, strip mining, Chernobyl, the greenhouse effect, and nuclear waste; civilization seemed to be drowning in the excrement of its own energy production. Cold fusion promised to liberate Earth from this slow strangulation.

The reaction of the scientific community to the news out of Salt Lake City was in sharp contrast to the indifference that had greeted Joe Newman's claim of unlimited energy five years earlier. Pons, after all, was a full professor of chemistry, with a lengthy record of research publications. Fleischmann, a visiting professor at Utah, was a professor at the University of Southampton and a member of the British Royal Society, a mark of considerable scientific distinction. They could not be ignored.

The day after the press conference, scientists in laboratories around the world were clustered around blackboards discussing the fragments of information that had appeared in the news. The first step was to ask whether the Utah claim was consistent with accepted physical principles. Initial calculations did not look promising. The information given to the press was, however, devoid of any details that might enable other scientists to judge the strength of the Utah claim or repeat the experiment. Calls to the University of Utah for more information produced nothing but the press release, which dealt more with the economic potential of cold fusion than with the scientific evidence.

This was no mere breach of etiquette. The integrity of science is anchored in the willingness of scientists to test their ideas and results in direct confrontation with their scientific peers. That standard of scientific conduct was being flagrantly violated by the University of Utah. Fleischmann and Pons were all over the news—

but they were not returning calls from other scientists. They had, as Joe Newman had, made their pitch directly to the media, and scientists were totally dependent on the media for information.

Extraordinary claims, as Carl Sagan said, are expected to be backed up by extraordinary evidence. An announcement of such importance would normally have been preceded by a careful review within the scientific community, and a detailed report would have been available to interested scientists at the time of the press conference. The basic claim of the two chemists, however, was clear: during electrolysis of heavy water (water in which ordinary hydrogen is replaced by deuterium), deuterium nuclei are squeezed so closely together in a palladium cathode that they fuse, releasing large amounts of energy. Deuterium is a naturally occurring stable isotope of hydrogen; its nucleus contains a neutron in addition to the single proton found in the nucleus of ordinary hydrogen. Since deuterium accounts for about 0.014 percent of all the hydrogen atoms in the ocean, the supply is inexhaustible.

The fusion of two deuterium atoms to form helium was studied by Ernest Rutherford at Cambridge as early as 1934. In the years since, few nuclear processes have been explored more thoroughly. Because of their positive charge, deuterium nuclei normally repel each other. If they can be forced close enough together, however, the short-range strong nuclear force takes over and the two nuclei "fuse" to form a compound nucleus consisting of two protons and two neutrons. These are the same particles that form the nucleus of ordinary helium, or helium-4, but while the helium-4 nucleus is very stable, the compound nucleus is created in a highly excited state, like an alarm clock that has been overwound. The spring snaps, and the compound nucleus violently casts off its excess energy.

The excess energy is carried away by nuclear radiation. About half the time, the radiation consists of a neutron ejected from the nucleus traveling at very high speeds. This converts the nucleus to helium-3, a lighter but stable isotope of helium. Both the helium and the energetic neutrons are unambiguous evidence that fusion has taken place. Because neutrons carry no electric charge, they easily escape through the walls of the experimental cell and can be detected with the right equipment.

Even at the highest concentrations of deuterium in the palladium cathode, however, it did not appear that deuterium nuclei would ever be close enough together to fuse. Fleischmann had explained in a television interview that the high concentration of deuterium in palladium was equivalent to a very high pressure. "Physicists," he said condescendingly, "have concentrated on high temperature; no one thought of using high pressure."

I was surprised by his comment; high concentrations of hydrogen isotopes in metals had been studied for years. Indeed, hydrogen isotopes in metals such as titanium, scandium, and erbium can reach concentrations two to three times greater than is possible in palladium. These metals are even used to store deuterium and tritium (a radioactive isotope of hydrogen with two neutrons in the nucleus) in certain components of nuclear weapons, and they are perfectly stable. Normally, when scientists think they have a new idea, the first thing they do is head for the library to see if someone else thought of it first. How, I wondered, could Pons and Fleischmann have been working on their cold fusion idea for five years, as they claimed, without going to the library to find out what was already known about hydrogen in metals?

There were other problems: the by-products of deuterium fusion, in addition to helium, are neutrons, tritium, and gamma rays. At the power levels claimed by Pons and Fleischmann, their test cell would be expected to emit lethal doses of nuclear radiation. Yet here were the two beaming chemists, in a photograph that appeared on the front pages of newspapers around the world, in jackets and ties, proudly holding their cell up for the cameras. As nuclear physicist Frank Close commented, it should have been the hottest source of radiation west of Chernobyl. There was skepticism even in the Physics Department at the University of Utah, where a dark joke making the rounds asked: "Have you heard the bad news about the research assistant in Pons's lab? He's in perfect health." If this was fusion, it was fusion of a sort never before seen, in which the energy is carried off as heat with little radiation. If it could not be explained by nuclear physics, the two chemists seemed to be saying, so much the worse for nuclear physics.

A few days after the Salt Lake City press conference, I was interviewed by NBC News chief science correspondent Robert Bazell. NBC, alone among the major news organizations, had not covered the story yet. Bazell had found a widespread conviction among scientists, particularly physicists, that the Utah claims were wrong. But no one seemed willing to say so bluntly on camera, and he was concerned that the public was not getting an accurate picture.

The interview took place in my Washington office. I briefly summarized why the cold fusion claim had to be wrong. As the NBC cameraman was packing up his equipment to leave, he asked me the question that couldn't be asked on camera: "So what's going on, doc? Is this fraud?" "I don't think so," I replied, "but give it a few weeks." Pons and Fleischmann had become world celebrities overnight, but in making their claim so forcefully and so publicly, they had left themselves no room to back down. If their claim did not hold up, and I was sure it would not, their character would be tested.

One reason Pons and Fleischmann had to be wrong was because the number of neutrons they claimed to see was at least a million times too small to account for the energy they reported. Still, if the experiment produced any neutrons at all, it would be proof that a nuclear process of some sort was taking place.

What makes such measurements difficult is a "noise background" of neutrons from cosmic rays. It's like trying to carry on a conversation at a noisy cocktail party; a hearing aid doesn't help, because it will also amplify the background chatter. You must stand as close together as possible and speak directly into each other's ear. At Yale, Moshe Gai and his students had built a large-aperture neutron detector that worked on the cocktail party principle: capture as much of the "signal" as possible. They were using the detector to measure very low levels of neutron emission from certain rare decays of atomic nuclei.

For the cold fusion studies, Gai had an inspiration: he would surround the cold fusion cell with two layers of detectors and count only those neutrons detected by both layers in the right sequence. In the cocktail-party analogy, it was like having a filter that would only let through one voice. He was so excited by the idea that he ate and slept in the nuclear physics lab for a month while he perfected the new detector. Eventually he would reduce the background level to an incredible two neutrons per day—which he jokingly named Stanley and Martin. Gai was confident that with such a detector he would be able to test the cold fusion claim, but he knew nothing about electrochemistry.

The head of the Yale nuclear physics lab was Allan Bromley. One of the most prominent scientists in the country, the politically conservative Bromley was rumored to be on the short list for the position of science advisor to newly elected president George Bush. Bromley realized that Gai would need an expert electrochemist as a collaborator. He arranged for Gai to meet Kelvin Lynn, a forty-one-year-old electrochemist at Brookhaven National Laboratory, just two hours away on the eastern end of Long Island. Lynn, it turned out, had been a student at the University of Utah and knew both Pons and Fleischmann. He was inclined to believe there must be something to their claim, and he was already starting to build the sort of electrolysis cell they used. It was a healthy collaboration: Gai and Lynn approached the problem with sharply different expectations.

Gai and Lynn were not the only ones obsessed with cold fusion. Scientists who had paid no heed to Joe Newman's Energy Machine were suddenly working day and night on cold fusion, even though the Utah claim seemed to violate much of what was known about nuclear physics. "If everybody knows it's wrong," a puzzled reporter asked me, "why are they all doing it?" Certainly there was the lure of possible new science, but for many there may also have been the smell of blood. "The crowd cheers just as loudly when the quarterback gets sacked as when he throws a touchdown," I told the reporter. But his question was troubling. Whatever their motives, scientists from all over the world, in every branch of physics and chemistry, were joining the race to test the Utah claim. It was starting to look more like a stampede.

Much the same sort of stampede had followed the discovery two years earlier of high-temperature superconductivity, another quite unexpected result that ran counter to many preconceived notions. There was one major difference: the discoverers of high-temperature superconductivity, Georg Bednorz and Karl Mueller at IBM's Zurich laboratory, who would share the Nobel Prize for their discovery a year later, adhered meticulously to the "rules" of scientific exchange. The public announcement of their discovery coincided with the publication in a peer-reviewed journal of every detail of their experiment. Everyone who tried it got the same result. Within weeks, high-temperature superconductors were being fabricated in junior high school science classes.

Any scientist who wanted to repeat the Pons-Fleischmann experiment, however, would first have to figure out what it was. Many scientists actually seemed to enjoy the challenge of penetrating the wall of secrecy thrown up by the Utah chemists. Overnight, an informal intelligence network took form. The Internet did not yet exist, but a computer bulletin board was set up to share information among laboratories all over the world. Fax machines were used to exchange press clippings. Video tapes of news interviews with Pons and Fleischmann were played over and over. Press photographs were blown up to get a closer look at the Utah apparatus; the width of Pons's wrist was used to estimate dimensions. Within days, other laboratories believed they were ready to repeat the cold fusion experiments of Pons and Fleischmann.

Initial press stories looked good for Utah. The scientific stampede set off by cold fusion was interpreted by many reporters as evidence that there must be something to it, and cold fusion was in the news almost daily. The mania fed on itself. Optimistic press reports encouraged other groups to go public with their own premature findings. A loose wire, electrode contamination, calibration errors, quirky detectors—all got reported as "anomalous results" or "partial confirmations."

When reports came in from groups that found no evidence of cold fusion, Pons and Fleischmann would explain that it took several days to "load" the cathode with deuterium before the reaction could begin. When these groups still saw nothing after a week, the Utah scientists said it sometimes took ten days; or three weeks; or the cathode was the wrong size; or they weren't using the right electrolyte; or it only worked with cast palladium and not with extruded. The rumor spread that Pons and Fleischmann were deliberately misleading researchers to conceal "the secret" while they negotiated with potential investors.

Nevertheless, on April 12, at the annual American Chemical Society meeting in Dallas, seven thousand chemists greeted Pons with a standing ovation. It was the first time since the March 23 press conference that the reclusive Pons had exposed himself to questions from other scientists. He shared the platform with physicist Harold Furth, the head of a laboratory at Princeton that had spent hundreds of millions of dollars attempting to create a sustained fusion reaction at high temperatures. It was this work that Pons had derided at the Salt Lake City news conference in his comment about "conventional techniques."

We should take a moment to talk about what "conventional" fusion is—or at least might be. It has been the dream of scientists since the 1950s to build a practical fusion reactor using the same principle that powers the Sun: a high-temperature plasma in which the nuclei of hydrogen isotopes would collide with one another so violently that they would overcome their mutual repulsion and fuse. In physics, the term *plasma* refers to a gas at such high temperature that electrons are stripped from the atoms, so that rather than being a gas of neutral atoms, a plasma is a gas of positively charged ions and negative electrons. But even on the Sun, fusion is a slow process. That's why the Sun does not simply explode. It has been "burning" for billions of years and will continue for many billions more before its fuel is exhausted.

It's easy enough to duplicate the process in the laboratory; the problem is that on a laboratory scale it isn't very efficient. A hot gas will try to expand, cooling as it does; the plasma must be confined in some way, and therein lies the problem. On the Sun, which thankfully burns rather slowly, the attempt of the hot gas to expand is balanced by the powerful gravity. In a practical fusion reactor, the plasma must be even hotter than the Sun. But what container on Earth could withstand such high temperatures?

Scientists have been trying to use a magnetic bottle. It works like this: a charged particle moving in a magnetic field experiences a force at right angles to its motion, causing it to spiral about the direction of the magnetic field in a helical, or corkscrew, trajectory. The stronger the magnetic field, the tighter the spiral. These helical paths behave like the windings of an electromagnet, generating their own field, which adds to the external field, further compressing the hot plasma and raising the temperature still higher. Using this "pinch effect," it was thought, a magnetic bottle could contain a plasma at the temperature needed for sustained fusion. When the pinch effect was demonstrated in the late 1950s, it was believed that fusion power plants were just around the corner.

Of course, the plasma was still free to leak out the ends of the cylinder. That was to be solved by bending the cylinder into a torus, or doughnut shape, so it would have no ends. Everything seemed to work as planned, until they tried to scale things up to the point where more energy could be produced by fusion than was needed to heat the plasma. Scale-up often produces surprises. The first automobile, for example, could travel about 10 miles per hour. Straightforward improvements in engines, suspensions, steering, drive trains, etc., have increased that by an order of magnitude (i.e., a factor of ten) to about 100 mph. However, any attempt to scale up another order of magnitude, to 1,000 mph, runs into the sound barrier at about 740 mph. It's still possible, but physical principles that could be ignored at 100 mph make it extraordinarily difficult.

So it was with fusion. The fusion reactor must pass break-even, the point at which the fusion energy exceeds the energy needed to heat the plasma. But before break-even could be achieved, unexpected instabilities set in, causing the magnetic bottle to spring leaks. It's a little like trying to compress a balloon between your hands: it begins to bulge out between your fingers. As ways are found to overcome one sort of instability, a new one seems to be waiting at somewhat higher temperatures. Steady progress has been made in overcoming these problems, but a practical magnetic-confinement fusion reactor seems farther in the future than ever. The capital costs may turn out to be so high that fusion will not be practical until most other sources of energy are depleted. Cynics scoff, "Fusion is the energy source of the future—and always will be."

The world's most advanced facility for magnetic-confinement fusion research was the Princeton Plasma Physics Laboratory, headed by Harold Furth. The centerpiece of the Princeton Laboratory was a mammoth toroidal confinement device of incredible complexity called a tokamak. Pons showed a slide of the Utah experiment—a tabletop arrangement set up in a Rubbermaid dishpan to provide a constant-temperature bath. "This," he deadpanned, "is the U-1

Utah tokamak." The huge audience roared with laughter. Unruffled, Harold Furth asked a single question: "What happens in your experiment if ordinary water is substituted for heavy water?"

Much of experimental research is a matter of designing "controls" to ensure that your results show what you think they do and not some flaw in the equipment or in the design of the experiment. A control experiment is intended to be as nearly identical to the real experiment as possible—except for one critical element. In this case, the water was just such a factor. Since the hydrogen atoms in ordinary water have no neutrons, they cannot directly fuse to form helium, which needs either one or two neutrons in the nucleus. If something you have been attributing to deuterium fusion is observed with ordinary water, it means you've been fooling yourself.

Pons replied that they had not tried using ordinary water, but he agreed it seemed like a good idea. How could the Utah chemists have been working on this for five years as they claimed without performing such an elementary control experiment? This tiny ripple on an ocean of optimism failed to warn reporters, and many scientists as well, that something was wrong beneath the surface. Many of the reporters interpreted Furth's question as sour grapes, and the press began to play the theme of a war between physicists and chemists. It made a good story, but at many laboratories, physicists and chemists—for example, the Moshe Gai–Kelvin Lynn collaboration—had already joined forces to tackle the problem of replicating the Utah experiment.

Pons returned to Utah from the meeting in Dallas, seemingly eager to try the "light-water" control experiment suggested by Furth, but a few days later when he was asked by a reporter what the result had been, Pons's only comment was a muttered, "We did not get the baseline we expected." Apparently the experiment behaved about the same with ordinary water as it had with deuterated water. Pons and Fleischmann would never mention the light-water experiment again.

This was a critical point in the evolving story of cold fusion. Pons had performed a critical test of their hypothesis—and it failed. But rather than accept the obvious conclusion, they chose either to ignore the result or to believe that somehow fusion could take place with ordinary hydrogen. Had the wish now taken over completely, erasing a lifetime of scientific training? Or was it something else?

Meanwhile, in New Haven, Allan Bromley had received a call from the White House. President George Bush, a Yale graduate, wanted to meet with him immediately to discuss the position of science advisor. Bromley was the natural choice for the job. He had served President Reagan on the White House Science Council and advised Bush during the campaign. Bromley fit the role perfectly. Physicists tend to be a lean and scruffy-looking lot, too absorbed in their work to pay much attention to their appearance. They don't fit comfortably into the dark suits and white shirts favored in Washington. Bromley, by contrast, was an impressive figure, given to conservative suits and bow ties, with an abundant shock of wavy white hair that curls upward on the back of his neck, and the sort of voice you imagine for a Roman senator. In any gathering, it always looked like he was in charge.

Bromley drove to La Guardia and caught the Eastern Shuttle to Washington. At the White House, the first question he was asked by President Bush was whether the cold fusion reports coming out of Utah were believable. Both the administration and Congress were under growing pressure to invest heavily in what was already being touted as the energy source for the twenty-first century. Bromley was ready for the question; the Yale-Brookhaven collaboration of Gai and Lynn had just briefed him on their preliminary findings. There was no neutron emission. He confidently informed the president that the reports out of Utah were in error.

In Utah, the state legislature, unaware of growing skepticism within the scientific community, met in special session to vote the University of Utah $5 million to begin developing cold fusion. The university said agreements would be signed only with companies that would base some of their effort in Utah. It was reported that James Fletcher, the retiring head of NASA and a former president of the University of Utah, had agreed to oversee the Utah cold fusion effort. Fletcher was a devout Mormon, and many Utah Mormons were convinced that the discovery of cold fusion came directly from God to rescue the state from serious economic problems.

At the Lawrence Livermore National Laboratory in California,

Edward Teller, the aging "father of the H-bomb" and an almost mythic hero to conservatives, had declared soon after the Salt Lake City press conference that cold fusion "sounds right." His protégé, Lowell Wood, anxious to prove his mentor correct, attempted to reproduce the Fleischmann-Pons experiment. Unfamiliar with electrochemistry, Wood set off an explosion in his laboratory when hydrogen, liberated by electrolysis, ignited. The blast shattered his apparatus and ended his quest for cold fusion.

In Washington, the Strategic Defense Initiative began organizing a workshop to examine whether cold fusion could power a Star Wars missile defense, and Senator Jake Garn of Utah was arranging for an Air Force transport to carry a load of Washington dignitaries to Salt Lake City for a firsthand look. Representative Robert Walker, the ranking Republican on the House Science Committee, submitted a budget amendment transferring $5 million from the Energy Department's program in conventional "hot fusion" to cold fusion, and the committee announced it would hold hearings on "recent developments in fusion energy." Cold fusion seemed to have pushed the rest of the scientific enterprise aside.

In New Haven, at 7:00 A.M. on April 21, Allan Bromley was in the shower when the phone rang. His wife Pat answered the phone. "Could I speak to Allan?" the familiar voice said. "This is George Bush." Unadorned and dripping water across his study, Bromley hurried to the phone. A puddle collected at his feet as the president asked him to take on the duties of science advisor. He was needed in Washington.

During this period, the public heard little about the growing skepticism of scientists. It is not, of course, up to the media to decide what is good or bad science. The media was reporting what it heard from scientists. Only a tiny fraction of all scientific research is ever covered by the popular media, however, and most scientists go through their entire career without once encountering a reporter. New results and ideas are argued in the halls of research institutions, presented at scientific meetings, published in scholarly journals, all out of the public view. Voodoo science, by contrast, is usually pitched directly to the media, circumventing the normal process of scientific review and debate. We saw this in the case of the Newman Energy Machine, the Patterson cell, and the cold fu-

sion claims of Pons and Fleischmann. The result is that a dispro-
portionate share of the science seen by the public is flawed.

The reluctance of scientists to publicly confront voodoo science
is vexing. While forever bemoaning general scientific illiteracy, scientists suddenly turn shy when given an opportunity to help ed-
ucate the public by exposing some preposterous claim. If they comment at all, their words are often so burdened with qualifiers that
it appears that nothing can ever be known for sure. This timidity
stems in part from an understandable fear of being seen as intol-
erant of new ideas. It also comes from a feeling that public airing
of scientific disputes somehow reflects badly on science. The result
is that the public is denied a look at the process by which new
scientific ideas gain acceptance. We will discuss that process in the
next chapter.

More perplexing was the overreaction of the scientific community to improbable claims based on the flimsiest of evidence and
shrouded in secrecy. The principal reason for keeping science secret,
after all, is that the science is questionable. Perhaps many scientists
found in cold fusion relief from boredom. Much of a scientist's
time is consumed with the routine labor of small advances; the
sudden insights—the "eureka moments"—that are the reward the
research scientist seeks can be separated by long periods of tedium.

We have left the cold fusion story unresolved, but we will return
to Pons and Fleischmann. They did not go gently. Nor have we
seen the last of Joe Newman and his Energy Machine, or James
Patterson and his magic beads. There are still lessons to be learned
from these episodes, and questions left unanswered. Was there—is
there—fraud involved? Did the scientific community rush to judgment in the spring of 1989? We will try to find answers to these
questions and look for parallels in other examples of voodoo science. But first we must ask why, faced with the same set of facts,
some believe and others doubt.

TWO
THE BELIEF GENE
In Which Science Offers a Strategy for Sorting Out the Truth

THE MOST COMMON OF ALL FOLLIES

IN 1995 THE NATURAL LAW PARTY succeeded in getting its presidential candidate, John Hagelin, on the ballot in all fifty states—a goal that had eluded other third-party hopefuls, even Ross Perot four years earlier. The platform of the Natural Law Party offered an "action plan to revitalize America," based on "scientifically proven solutions." The centerpiece of the scientific proof was an experiment conducted in Washington, D.C., in the summer of 1993.

More than five thousand experts in Transcendental Meditation (TM) from around the United States and eighty countries worldwide spent two-week shifts in the nation's capital as part of the National Demonstration Project to Reduce Violent Crime in Washington, D.C. Mostly young white professionals, they began arriving on June 5. Their objective in the coming weeks would be to meditate in unison, creating a "coherent consciousness field" that would produce a calming effect, not just among the meditators but throughout the city. Organizers of the $6 million project predicted that violent crime in the city would be reduced by 20 percent.

The head of the project was John Hagelin, a thirty-nine-year-old physicist with a receding hairline and a perpetual cherubic smile. His high forehead was unfurrowed by negative thoughts. A summa cum laude graduate of Dartmouth, Hagelin had gone on to complete a Ph.D. in physics at Harvard. In 1983 he was regarded as a competent theoretical physicist and had a postdoctoral research appointment at the Stanford Linear Accelerator; then, in the midst of personal problems, he simply vanished, reappearing a year later as chairman of the Physics Department at Maharishi International University in Fairfield, Iowa. The university was founded by the Maharishi Mahesh Yogi, the Indian guru who vaulted to fame after becoming the spiritual advisor to the Beatles.

Hagelin held a press conference in the District Building to announce the violence-reduction project. Once a beautiful example of a classic white-marble municipal building, the crumbling structure seemed to symbolize the inability of the District of Columbia to govern itself. The formerly broad halls had been narrowed by ramshackle partitions erected to create more offices for political appointees. In a conference room with paint peeling from the walls, Hagelin explained that the Project to Reduce Violent Crime was a "scientific demonstration that will provide proof of a unified superstring field." Superstring theory is an abstract and highly speculative physical theory that attempts to connect all the forces of nature. According to Hagelin, one such force is a collective consciousness that can be accessed by TM. A superstring field, generated by many minds meditating in unison, would radiate throughout the community, reducing stress and spreading tranquility.

The weeks that followed seemed like something out of an old mad-scientist movie—an experiment that had gone horribly wrong. Each Monday morning, the *Washington Post* would tally the gruesome weekend slayings in the city. Participants in the project seemed serenely unaware of the mounting carnage around them as they sat cross-legged in groups throughout the city, eyes closed, peacefully repeating their mantras. The murder rate for those two months reached a level unmatched before or since.

At the end of the demonstration period, Hagelin, smiling his unworldly smile, acknowledged that murders were indeed up "due to the unusually high temperatures," but "brutal crime" was down. One could only imagine that the murders were being committed more humanely—perhaps a clean shot between the eyes rather than a bludgeoning. Over the coming year, Hagelin promised, the results would be carefully analyzed according to strict scientific standards.

As promised, Hagelin was back a year later with a fifty-five-page report of the results of the project. It was a clinic in data distortion. A beaming Hagelin announced at a press conference that, during the period of the experiment, violent crime had been reduced by a remarkable 18 percent. "An eighteen-percent reduction compared to what?" a puzzled reporter for the *Washington Post* asked, recalling the dreadful murder rampage of the summer of '93. Compared to what it would have been if the meditators had not been meditating, Hagelin explained patiently. "But how could you know what the rate would have been?" the reporter persisted. That had been arrived at, Hagelin responded with just a trace of irritation, by means of a "scientifically rigorous time-series analysis" that included not only crime data but such factors as weather and fluctuations in Earth's magnetic field.

According to Hagelin, their analysis showed a significant reduction in psychiatric emergency calls, fewer complaints against the police, and an increase in public approval of President Clinton during the period of the experiment—all consistent with the hypothesis that a coherence-creating group of TM experts can relieve communal stress and reverse negative social trends. All of this had been carefully scrutinized by an "independent scientific review board," several of whose members were present at the press conference. Hagelin was clearly irritated when I asked how many of the "independent" review board members practiced TM. "Some members of the review board have had previous experience with TM," he replied, struggling to retain some trace of his smile. He lost the struggle when I insisted on polling the members of the scientific review board. They were all followers of the Maharishi.

"The most common of all follies," wrote H. L. Mencken, "is to believe passionately in the palpably untrue." The belief of the Maharishi's followers in the power of TM was not influenced in the slightest by the outcome of the "experiment." This was pseudoscience: all the talk of "string theory" and "consciousness fields" and "time-series analysis," was meant to give the appearance of science. Which is not to say that those involved were not sincere in their belief. They may have believed so fervently that they felt a responsibility to *make* the facts support their belief. People will work every bit as hard to fool themselves as they will to fool others—which makes it very difficult to tell just where the line between foolishness and fraud is located.

The vast majority of scientific research, of course, is far removed from either foolishness or fraud. But to what extent are the interpretations given to scientific evidence shaped by the worldview of the scientist? A good place to examine this question is the current controversy over global warming.

THE GREAT GLOBAL WARMING DEBATE

André Gide, the great French moralist, wrote in his journal a half century ago: "Man's responsibility increases as that of the gods decreases." Every step taken by science claims territory once occupied by the supernatural. Where once we accepted storms and drought as divine will, there is now overwhelming scientific evidence that we ourselves can affect Earth's climate. It is a measure of how far science has come that scientists have been given responsibility for telling us whether our planet is headed for some climate catastrophe of our own making, and if so, what steps we can take to avoid it.

The evidence comes from a revolution in climate research over the past decade, brought about by new observational techniques, including satellites, and a prodigious increase in computational and data-storage capabilities made possible by microelectronics. It now seems undeniable that surface temperatures are warmer than they were a hundred years ago. There is also no doubt that the burning of fossil fuels since the beginning of the industrial revolution has resulted in a significant increase in atmospheric carbon dioxide.

What is in dispute is what the long-term consequences of continued carbon dioxide increases will be for Earth's climate and the quality of life. Carbon dioxide, or CO_2, is called a "greenhouse gas," because like a greenhouse, or your car when it's parked in the sun with the windows closed, it traps heat. Some fraction of the sunlight that strikes the Earth is absorbed, warming the planet, which then reradiates energy. But because it is not nearly as hot as the sun, whose light is most intense in the yellow-green region of the visible spectrum, the Earth radiates at much longer wavelengths, peaking in the invisible infrared region of the spectrum. CO_2, like glass, is transparent to the rays of visible sunlight that warm the Earth but blocks heat from radiating back into space. The presence of CO_2 and other greenhouse gases in the atmosphere helps to keep our planet warm. CO_2 is also the raw material for plant growth. Using the energy of sunlight, plants draw CO_2 from the air to make hydrocarbons, releasing oxygen into the atmosphere as a by-product. When the plant dies and decays, or is burned, or is eaten by an animal, the carbon is recombined with oxygen and returned to the atmosphere as carbon dioxide, completing the cycle.

Before the industrial revolution, the concentration of carbon dioxide represented a natural balance, but in a little more than a century, humans have disrupted that balance by burning fossil fuels that were built up in underground deposits over a period of hundreds of millions of years. If this release of carbon dioxide into the atmosphere continues, climatologists warn, there could be disastrous consequences in the next century: many of the world's great cities will be submerged by rising sea levels as the polar ice caps melt, and drastic changes in rainfall patterns could wreak havoc on food production.

The average temperature of the Earth has risen by perhaps one degree Fahrenheit in this century, and it would be more if we had not also polluted the atmosphere with soot, blocking out some of the Sun's rays. The greatest concern is that there are feedback mechanisms that might cause this gradual warming to accelerate. Thawing tundra, for example, would release trapped methane, another greenhouse gas, causing still more warming. Warming would also reduce the amount of sea ice. A large fraction of the sunlight that falls on ice is reflected back into space, but water absorbs sunlight rather efficiently. If the area of the Earth covered by sea ice shrinks, the warming will accelerate further. There is evidence of such rapid warming in prehistoric times. The nations of the world, many scientists argue, should take immediate steps to control the burning of fossil fuels, at least until we can better predict the consequences. We have no right, they declare, to place future generations in jeopardy.

Not all scientists agree. A number of prominent scientists point out that there were periods of global warming long before humans began burning fossil fuels, and CO_2 is a relatively minor greenhouse constituent in the atmosphere. They contend that any rise in global temperature since 1850 may simply be the result of natural solar variations. Some go further, describing the increase in carbon dioxide as "a wonderful and unexpected gift of the industrial revolution." The increase in atmospheric CO_2 has stimulated plant growth, making this a lusher, more productive world, capable of sustaining a much larger population. Besides, if there is some greenhouse effect, it may be just what Earth needs to stave off another ice age. The more industrial growth we have, including increased burning of fossil fuels, they argue, the better off we will be. They stop just short of telling us we have a moral obligation to burn more hydrocarbons.

If scientists all claim to believe in the scientific method, and if they all have access to the same data, how can there be such deep disagreements among them? If the climate debate was just about the laws of physics, there would be little disagreement. What separates the two sides in the climate controversy, however, is not so much an argument over the scientific facts, scientific laws, or even the scientific method. The climate is the most complicated system scientists have ever dared to tackle. There are huge gaps in the data for the distant past, which, combined with uncertainties in the computer models, means that even small changes in the assumptions result in very different projections far down the road. Neither side disagrees with that. Both sides also agree that CO_2 levels in the atmosphere are increasing. What separates them are profoundly different political and religious worldviews. In short, they want different things for the world.

The great global warming debate, then, is more an argument about values than it is about science. It sounds like science, with numbers and equations and projections tossed back and forth, and the antagonists believe sincerely that they are engaged in a purely scientific debate. Most scientists, however, were exposed to political and religious worldviews long before they were exposed in a serious way to science. They may later adopt a firm scientific worldview, but earlier worldviews "learned at their mother's knee" tend to occupy any gaps in scientific understanding, and there are gaps aplenty in the climate debate.

This sort of dispute is seized upon by postmodern critics of science as proof that science is merely a reflection of cultural bias, not a means of reaching objective truth. They portray scientific consensus as scientists voting on the truth. That scientists are influenced by their beliefs is undeniable, but to the frustration of its postmodern critics, science is enormously successful. Science works.

We will come back to our example of the climate war later in this chapter, but to understand how science can rise above the beliefs of its practitioners, we must first understand something of the process by which beliefs are generated.

PLEISTOCENE PARK

To borrow from the premise of the movie *Jurassic Park*, suppose a mosquito gorged on one of our Cro-Magnon ancestors thirty thousand years ago and then became trapped in amber, providing science with ancient human DNA. Would a Cro-Magnon clone, raised in today's society, be some dangerous brute that might escape and terrorize society? The movie *Pleistocene Park* would not be that exciting. A Cro-Magnon would most likely be indistinguishable from the rest of us. Far too little time has passed for any genetic adaptation to the modern world. All of recorded history covers a mere five thousand years—the industrial revolution just two hundred—the space age barely four decades. So here we are, saddled with stagnant genes that were selected for life as Pleistocene hunter-gatherers, trying to cope with a world of jet travel and computers. What provided a survival advantage in a Pleistocene wilderness, does not necessarily do so today.

Behavioral traits are as much a part of our genetic inheritance as physical characteristics. We respond to external stimuli in ways that conferred some sort of survival advantage on our distant human and prehuman ancestors. Psychologist James Alcock describes our brains as "belief engines," constantly processing information that comes in from our senses and generating new beliefs about the world around us. These new beliefs are selected by the brain to be consistent with beliefs already held, but they are generated without any particular regard for what is true and what is not.

A belief begins when the brain makes an association between two events of the form: B follows A. The next time A occurs, the brain is primed to expect B to again follow. The survival advantage of such a strategy for our primitive ancestors is obvious. They had scant means for separating causal connections from mere coincidence—better to take heed of every connection and be safe. We avoid some food, for example, because we once got sick after eating it. Our illness may have had nothing to do with the food, but unless we're facing starvation, there's not much to be lost by avoiding it.

Information gathered by the senses is normally routed through the thalamus, a small subsection deep within the brain, to the sensory cortex, which analyzes it in detail to decide how much weight it should be given. An exception is olfactory input, which apparently follows more evolutionarily ancient pathways to reach the cortex. Sensory information processed by the cortex finally reaches the amygdala, almond-shaped structures in the temporal lobes. The amygdalas contribute the emotional portion of our response to sensory stimuli. Parts of the amygdala, for example, are involved in fear. Animals with damage to these parts are no longer perturbed by stimuli they previously learned to fear.

Whether a belief is retained depends on how significant B is— how frightened we were, for example—and whether the association with A gets reinforced. Without reinforcement, the expectation that B will follow A will usually fade in time. If B again follows

A, however, it may still be a coincidence, but it will now be far harder to persuade us of that.

The belief may also be permanent if the information entering the thalamus coincides with a high state of emotional arousal, such as fear or the thrill of victory. The chemical messengers of emotion cause the thalamus to bypass the sensory cortex and route the information directly to the amygdala. This is often the origin of what might be called personal superstitions—the golfer who won't play without his lucky hat, for example. People develop elaborate rituals in an effort to re-create the conditions that surrounded some rewarding experience or to avoid conditions their brains associate with fear or pain. We often find ourselves almost compelled to go through these rituals, even when the cerebral cortex is telling us that a causal connection is highly implausible.

This kind of belief generation was going on long before our ancestors began to resemble humans, of course, but the advent of language opened a powerful new channel, both for the formation of beliefs and for their reinforcement. Speech exposes us to the generation of shared beliefs—beliefs based not on personal experience but on experiences related to us by others. This has the potential to spare us a lot of unpleasantness. Everyone, for example, need not discover the hard way that a particular plant is poisonous. The shared beliefs of a family or tribe are also a powerful force of social cohesion and are reinforced throughout our lives. Language makes vicarious experience the dominant source of belief in humans, overwhelming personal experience. The power of language was enormously amplified by the invention of writing and continues to be amplified by every new advance in communication from the printing press to the World Wide Web. Beliefs can now spread around the world in the twinkling of a computer chip. That which allows us to learn from others, unfortunately, also exposes us to manipulation by them.

Small children are particularly open to new beliefs, accepting without question whatever they are told by adults. Their belief engine runs freely, finding few previous beliefs to contradict what they are told. For a small child who must quickly learn that stoves burn and strange dogs bite, this sort of credulity is important to survival. Because a child's beliefs are not enmeshed in a network of related beliefs, however, children seem able to cast them off almost as easily as they adopt them. Fantastic stories about Santa Claus and tooth fairies, which are accepted uncritically, are dropped just as uncritically when someone, often a playmate, explains that it isn't really so. Nor do children appear to develop doubts about other things they've been taught, just because the Santa Claus story was taken back.

As the store of beliefs grows, conflicts with existing beliefs become more likely, and doubt begins to manifest itself. By the time the child reaches adolescence, beliefs tend to be enmeshed in an insulating matrix of related beliefs. The belief process becomes decidedly asymmetric: the belief engine is generating beliefs far more easily than it erases them. Once people become convinced that a rain dance produces rain, they do not lose their belief in years the drought persists. They are more likely to conclude that they have fallen out of favor with the Rain God, and perhaps add a human sacrifice to the ritual.

The result is that most of us wind up with beliefs that closely resemble those of our parents and community. Society, in fact, often holds it to be a virtue to adhere to certain beliefs in spite of evidence to the contrary. Belief in that which reason denies is associated with steadfastness and courage, while skepticism is often identified with cynicism and weak character. The more persuasive the evidence against a belief, the more virtuous it is deemed to persist in it. We honor faith. Faith can be a positive force, enabling people to persevere in the face of daunting odds, but the line between perseverance and fanaticism is perilously thin. Carried to extremes, faith becomes destructive—the residents of Jonestown for example, or the Heaven's Gate cult. In both cases, the faith of the believers was tested; in both cases, they passed the test.

The wonder is not that we can be easily fooled but that we function as well as we do on what would seem to be, as far as our genes are concerned, an alien planet that does not at all resemble the wild planet on which our genes were selected. If this sounds hopelessly gloomy, be patient, we are coming to the good news: we are not condemned to suffer the tyranny of the belief engine. The primitive machinery of the belief engine is still in place, but evolution didn't stop there. It provided us with an antidote.

WHAT IS SCIENCE?

How can it be that brains designed for finding food and avoiding predators in a Pleistocene forest enable us to write sonnets and do integral calculus? We invent poetry and higher mathematics because our brains hunger for patterns. The wonderful pattern recognition equipment residing in the higher centers of the human brain allowed our ancestors to adapt to changing conditions with remarkable ease, by quickly picking up the patterns that are characteristic of the new environment.

Animals with much smaller brains than ours also rely on pattern recognition, of course. The desert *Cataglyphis* ant, for example, whose brain contains perhaps a hundred thousand brain cells, compared to a million times that many for a human, forages over enormous expanses of seemingly featureless terrain, wandering to and fro in search of food. When these ants finally encounter some wind-blown seed, they return with it at once to their nest in an almost straight line. They navigate by the position of the Sun—even if the Sun is obscured by clouds—using patterns of polarized light. But the ability of *Cataglyphis* to recognize patterns, as marvelous as it is, is very specialized. Transplanted to a different environment, such as the forest floor, where landmarks abound but where the sky cannot be seen, *Cataglyphis* would be lost.

In humans, the ability to discern patterns is astonishingly general. Indeed, we are driven to seek patterns in everything our senses respond to. So far, we are better at it than the most powerful computer, and we derive enormous pleasure from it. Pattern recognition is the basis of all esthetic enjoyment, whether it is music or poetry or chess or physics. As we become more sophisticated, we seek out ever more subtle patterns. So intent are we on finding patterns, however, that we often insist on seeing them even when they aren't there, like constructing familiar shapes from Rorschach blots. The same brain that recognizes that tides are linked to phases of the moon may associate the positions of the stars with impending famine or victory in battle.

That is again the belief engine at work. But once we recognize how easily we can be fooled by the workings of the belief engine, we can use the higher centers of the brain to consciously construct a more refined strategy that combines our aptitude for recognizing patterns with the accumulation of observations about nature made possible by language. Such a strategy is called science.

> Science is the systematic enterprise of gathering knowledge about the world and organizing and condensing that knowledge into testable laws and theories.

This elegant description, borrowed from biologist E. O. Wilson's *Consilience*, provides a template that can be held up against claims to see if they belong in the realm of science. How well the template fits comes down to two questions: Is it possible to devise an experimental test? Does it make the world more predictable? If the answer to either question is no, it isn't science.

The success and credibility of science are anchored in the willingness of scientists to obey two rules:

1. Expose new ideas and results to independent testing and replication by other scientists.
2. Abandon or modify accepted facts or theories in the light of more complete or reliable experimental evidence.

Adherence to these principles provides a mechanism for self-correction that sets science apart from "other ways of knowing," to use a fashionable euphemism. When better information is available, science textbooks are rewritten with hardly a backward glance. Many people are uneasy standing on such loose soil; they seek a certainty that science cannot offer. For these people the unchanging dictates of ancient religious beliefs, or the absolute assurances of zealots, have a more powerful appeal. Paradoxically, however, their yearning for certainty is often mixed with respect for science. They long to be told that modern science validates the teachings of some ancient scripture or New Age guru. The purveyors of pseudoscience have been quick to exploit their ambivalence.

Scientists generally believe the cure for pseudoscience is to increase science literacy. We must ask, however, what it is we would want a scientifically literate society to know. There are a few basic concepts—Darwinian evolution, conservation of energy, the periodic table—that all educated people should know something about, but the explosive growth of scientific knowledge in the last half of the twentieth century has left the scientists themselves struggling to keep up with developments in their own narrow specialties. It is not so much knowledge of science that the public needs as a scientific worldview—an understanding that we live in an orderly universe, governed by physical laws that cannot be circumvented.

Although the old belief-generating machinery of the brain is still in place, habits of critical thinking can be adopted that subject each fledgling belief to skeptical analysis before continued reinforcement renders the belief hopelessly resistant. The first question that must be asked about a fledgling belief is whether *B* really follows *A* any more frequently than we would expect from chance. The belief engine, of course, knows nothing of the laws of probability. Any such analysis must be consciously imposed by the higher centers of the brain.

Most people, for example, will grant that a coin toss will come up heads or tails with equal probability. They will even concede that this must be true every time the coin is tossed. And yet, if the coin comes up heads four times in a row (which it has one chance in sixteen of doing), it takes a certain amount of mental discipline not to believe that the fifth toss is more likely to be tails. The part of our brain that understands that heads and tails are equally likely expects the tails to start catching up. This is known as the "gambler's fallacy." Heads and tails will tend to even out over the long run, but this says nothing about the next toss.

We must also ask if there is a plausible mechanism by which *A* could cause *B*. Even if we are satisfied that the connection between *A* and *B* is more than a coincidence, it still does not mean that *A* causes *B*. They could, for example, have a common cause. Ideally, we might know some physical principle that would help us decide, but more generally we have to decide whether that's the way other things seem to behave.

In 1934 the great chemist Irving Langmuir, who won the 1932 Nobel Prize for his studies of molecular films, read about the work of Duke University psychologist J. B. Rhine on extrasensory perception (ESP). Langmuir was fascinated by what he called "pathological science—the science of things that aren't so." Its practitioners, he argued, are not dishonest; they simply manage to fool themselves. To Langmuir, ESP appeared to be a classic example of pathological science.

Among the symptoms that Langmuir associated with pathological science was that the evidence always seems to be at the very limit of detectability. In our cocktail party analogy it would mean you could just barely make out what was being said over the din of background noise. Under these conditions it's easy to be mistaken about what was said.

If the claim is that the mind can influence the toss of a coin, for example, the reported success rate might be 51 percent rather than the 50 percent you would predict. Thus a great many trials would be needed to be reasonably sure that such a small deviation from pure chance is anything but expected random variation. But now there is a new problem: if there is some systematic flaw in the design of the experiment—perhaps some slight asymmetry in the two sides of the coin that influences which side is more likely to come up—it would produce a noticeable result only after a large number of trials. One experimentalist measuring a 51 percent success rate after a very large number of trials might therefore conclude that the result reveals the existence of an unidentified flaw in the experimental design and seek to identify the flaw. Another might conclude that the subject was able to mentally influence the coin and not look for flaws. Scientific claims that are based on small statistical differences, therefore, always carry less weight.

Another common characteristic of pathological science, Langmuir observed, is that there seems to be no way to increase the magnitude of the effect. To hear a sound more clearly, as in our discussion of conversations at cocktail parties in the last chapter, you move closer to the source, but neither distance nor time seemed to affect ESP. It didn't matter if the coin was tossed in some other city; the success rate would be the same. That, Langmuir pointed out, is certainly contrary to the way everything else in the world seems to work.

If the success rate was truly greater than chance, however, no matter how slight the advantage, it would be a profoundly important result, forcing a complete reexamination of all our assumptions about the way the world works. Langmuir visited Rhine and explained his reservations. To his surprise, Rhine seemed unperturbed and even urged Langmuir to publish his views. The result, Rhine predicted, would be that his ESP research would attract more graduate students and more funding. Moreover, Rhine was quite open about showing Langmuir how he conducted and analyzed his experiments.

Rhine had carried out hundreds of thousands of trials over the years involving the ability of people to guess the identity of cards dealt face down. He used a deck with five different cards, and in each trial the subject would be asked to guess the identity of twenty-five cards. On average you would expect people to guess correctly 20 percent of the time, thus getting five of the twenty-five right. Sometimes, of course, the subject would score better than five, other times worse. But out of a huge numbers of trials, Rhine found, the average was somewhat greater than you would predict by chance.

To his amazement, however, Langmuir discovered that in calculating his averages Rhine left out the scores of those he suspected of deliberately guessing wrong. Rhine believed that persons who disliked him guessed wrong to spite him. Therefore, he felt it would be misleading to include their scores. How did he know they deliberately guessed wrong? Because their scores were too low to have been due to chance. Indeed, he was convinced that abnormally low scores were as significant as abnormally high scores in proving the existence of ESP.

When Langmuir attempted to explain the flaw in Rhine's reasoning to a reporter, the reporter was unable to follow Langmuir's statistical arguments. What he wrote was that a famous Nobel laureate was looking into ESP. Rhine was overwhelmed with new graduate students and offers of financial support. As Rhine had expected, Langmuir had given ESP credibility simply by taking notice of it.

This creates a troubling dilemma for scientists. Joe Newman's challenge to scientists to debate him may have been rhetorical, but had some prominent physicist taken up his challenge, it would almost certainly have worked to Newman's advantage. Simplistic arguments and homespun humor are more effective in such a debate than citing the laws of thermodynamics. Debate has a way of seeming to elevate a controversy into an argument between scientific equals. It is an arena made for voodoo science.

The final step in applying a scientific worldview is to put a fledgling belief to the test. When I was young boy interested in nature, I read in one of my books that raccoons always wash their food before eating. I had in fact been told the same thing by my father, and I had even seen raccoons swishing their food in the edge of a stream, so there was not much reason to doubt it. The book explained that this behavior was not actually meant to cleanse the food but only to moisten it, because raccoons have no salivary glands. It seemed to be a reasonable explanation, and I carried this bit of lore around in my head for most of my life, eventually passing it on to my own children.

One summer, however, during a period of prolonged drought, a family of hungry raccoons began coming up to our house every evening at dusk to beg for food. They were impossible to resist, and we began buying dry dog biscuits for them, which we kept in a shed behind the house. Because the poor raccoons had no salivary glands, I would put out a pan of water first so they could moisten the food. They would crowd around me as I opened the shed and took out the paper bag of dog biscuits. Very soon, however, I noticed that at the first rattling of the paper bag, the raccoons would start salivating — saliva literally dripped from their jaws. No salivary glands indeed! After that, I tried feeding them without the pan of water. It didn't seem to bother them; they ate anyway. If the water was there, they used it. If it wasn't, they went right ahead and ate. I still don't know why raccoons like to swish their food in the water. My guess is they're washing it. The lesson is that no matter how plausible a theory seems to be, experiment gets the final word.

BACK TO THE CARBON DIOXIDE WAR

Which brings us back to the global climate-change debate. The special responsibility of scientists is to inform the world of its choices. During some three and a half billion years of evolution, the environment shaped our genes. Our genes are now shaping the environment. But it may be years before anthropogenic effects on climate are well enough understood to make those choices clear. On one side, there are scientists who warn that we can't afford to wait. These Malthusian pessimists argue for the "precautionary principle." Changing human behavior takes time, they contend, and if we don't start now it may be too late to prevent a catastrophe.

On the other side are the technological optimists, who insist that to make policy before we understand the problem, if indeed a problem exists, is to to invite failure. To have followed such a policy in the past, they argue, would have denied the world the unquestioned benefits of industrialization. They remind us that science has always found solutions to the problems generated by population growth and industrialization.

In the spring of 1998, a research group (group A) analyzing data from weather satellites concluded that over a twenty-year period there has been a slight cooling of the upper atmosphere, rather than the slight warming inferred from surface measurements. However, a second group (group B) reexamined the data and pointed out that the analysis failed to take atmospheric drag into account. That would put the satellite trajectory fifteen kilometers closer to Earth, which had the effect of turning slight cooling into slight warming. Group A thanked group B for pointing out the correction but were led thereby to reexamine the data themselves. They found that two further corrections, for orbital precession of the satellites and calibration drift in the radiometer, largely offset the effect of atmospheric drag. Group B appreciated this latest refinement but felt these effects were too small to change the conclusion that the troposphere is warming.

The most significant lesson from the satellite data may be that the ideological passion of Malthusian pessimists at one extreme and technological optimists at the other—so long as both sides adhere to the scientific process—actually serves as a powerful motivation for better climate science. Each side knows that every flaw in their data or oversight in their analysis will be seized upon by their opponents. Both sides strive to produce better data and better analysis in the conviction that the truth will favor their prejudice. The numbers, when science finally learns them, will ultimately decide the winner. In the end, the result will be a better understanding of global climate.

Of the multitude of problems that daily vex modern society, few, it seems, can be sensibly resolved without recourse to the knowledge of science. There are times, however, when society cannot wait for the scientists to get it right. The courts must resolve disputes, Congress must enact legislation, government agencies must impose regulations, doctors must treat the ill, all on the basis of the best scientific evidence available at the time. There no longer seems to be any reasonable doubt that human activity is affecting Earth's climate. Governments must initiate some precautionary measures, even though the precise consequences are still unclear.

The need to make decisions involving scientific questions that are as yet unresolved creates an inevitable tension between those who mistrust technology and those who trust it too much. At these two extremes, the scientific process is sometimes circumvented, giving rise to voodoo science, as we will see in the next two chapters.

THREE
PLACEBOS HAVE SIDE EFFECTS
In Which People Turn to "Natural" Medicine

CODFISH THRIVE ON IT

THERE WAS A FULL-PAGE AD in *USA Today* recently for "Vitamin O." Beneath a photograph of an attractive group of vigorous, smiling people, the ad said "Vitamin O" was helping thousands of people to live healthier lives. "It's so safe you can drop it in your eyes, so natural it contains the most abundant element on earth, so effective you could spend hours reading the unsolicited testimonials of those who've used it with dramatic results." Indeed, the ad included a number of testimonials, which said things like: "After taking 'Vitamin O' for several months, I find I have more energy and stamina and have become immune to colds and flu."

"Vitamin O" was to be taken orally as a supplement. The recommended dose was fifteen to twenty drops two or three times a day. According to the ad, it "maximizes your nutrients, purifies your bloodstream, and eliminates toxins and poisons—in other words, all the processes necessary to prevent disease and promote health." A two-ounce vial, which sold for twenty dollars plus shipping charges, should last a month. According to a company official at Rose Creek Health Products, which marketed "Vitamin O," sales were running about sixty thousand vials per month and growing.

So what *is* "Vitamin O"? The ad says exactly what it is: "stabilized oxygen molecules in a solution of distilled water and sodium chloride." In other words, it's salt water. There is always some oxygen dissolved in water, although at room temperature and atmospheric pressure the solubility is extremely small, amounting to no more than 7.5 ppm (parts per million). Higher solubility would require sealing the container under pressure. But it really doesn't matter how high the solubility is. At any solubility, the recommended dose would provide far less oxygen than you get in a single deep breath. Oxygen inhaled into the lungs goes directly into the bloodstream, and even at rest we take a breath about every three seconds.

Rose Creek Health Products seemed to have correctly gauged the sophistication of the public. The ad may be a cynical and apparently successful attempt to mislead, but for the most part it is literally truthful. It says the product is safe. What could be safer? It says oxygen is good for you. You certainly can't live without it. It says "Vitamin O" provides you with oxygen. I suppose it does—but in an amount that is totally insignificant. Fish can extract the oxygen they need from water, but they don't get it by swallowing water. Their gills are constantly processing water to extract dissolved oxygen. Moreover, fish are cold-blooded creatures with a relatively low metabolic rate. No warm-blooded creature survives without breathing air. An attempt to extract the oxygen you need from water is called "drowning."

"Vitamin O" is meant to appeal to people who feel left behind by the scientific revolution. Paradoxically, however, their nostalgia for a time when things seemed simpler and more natural often competes with a grudging respect for science. They want to believe that natural medicine and science will converge.

This ambivalence is cleverly exploited; the ad falsely claims that pollution and deforestation have reduced the concentration of oxygen in the atmosphere, particularly in major cities. Beside an image of an astronaut floating in a space suit, "Vitamin O" is described as the "newest generation of super oxygenation technology developed by Dr. William F. Koch for use by the astronauts to insure that they received enough oxygen to maintain their health." Reliance on Nature is mingled with the space age.

A century ago in the United States, worthless patent medicines, promoted by hyperbolic claims of cures for every ailment of mankind, were marketed without restraint. Public outrage eventually resulted in passage of the 1906 Pure Food and Drugs Act. Although the law prohibited false and misleading claims and allowed publication of the results of federal investigations, it provided no true enforcement powers. But for most of the rest of the century, Congress expanded the power of the federal government to protect citizens from dangerous or ineffective products.

The Food and Drug Administration (FDA), which was formed as a separate law enforcement agency in 1927, was empowered to prevent untested products from being sold and to take legal action to halt the sale of unsafe products. The thalidomide tragedy in the 1950s, which resulted in the birth of thousands of babies with missing or deformed limbs, led to additional amendments to the Food and Drug Act, requiring drugs to be proven safe and effective before they could be marketed. Most people assume that these laws prevent companies from making deliberately exaggerated and misleading claims for health products. Unfortunately, that is no longer the case.

In the 1990s, Congress began to turn the clock back. The Dietary Supplement and Health Education Act of 1994, passed in response to a huge lobbying campaign by the supplement industry, exempted "natural" dietary supplements from requirements for testing of safety, purity, or effectiveness. The FDA can go to court to get a dietary supplement taken off the market only if it can demonstrate that it's harmful—"not until the bodies start piling up," as one FDA official put it. No bodies will pile up as a result of taking "Vitamin O." It becomes dangerous only if it leads someone who needs real medical treatment to forego it.

The only legal requirement is that dietary supplements not be promoted as preventing or treating disease. Compliance has turned into a tightrope act. The Federal Trade Commission (FTC) has the authority to regulate the advertising, but it is often in the form of testimonials from satisfied users. If it's genuine, a testimonial is difficult to prosecute as false advertising, no matter how preposterous it may seem. The FTC, of course, has limited resources. Inundated with cases of false and misleading advertising, it can afford to take action only in the most serious cases. The FTC did not take action against Rose Creek until the "Vitamin O" scam was exposed in the media. I had called attention to it in a brief interview with *Science* magazine, and then on the National Public Radio program *All Things Considered*.

Two months later, on March 11, 1999, the FTC filed a complaint in U.S. district court against Rose Creek Health Products, the suppliers of "Vitamin O," charging the company with making blatantly false claims. In addition to the preposterous health claims, the FTC noted that "Vitamin O" was never given to astronauts as the ad implied. The FTC was unable to confirm that the "developer," Dr. Koch, even existed. Rose Creek Health Products was forced to cease advertising "Vitamin O," and the FTC has asked the court to compel the company to refund consumers' money. Millions of dollars are at stake.

But how could it be that so many people used "Vitamin O" and even testified to its effectiveness? If we are to believe the sales figures released by Rose Creek, there were more than sixty thousand people using "Vitamin O." Presumably they bought it because they thought it made them feel better. In the next section we will try to understand *why* they felt better.

DOES THE CROWING OF THE COCK MAKE THE SUN RISE?

The road to modern medicine is littered with the bones of medical treatments that millions of people once swore by—and are now known to be worthless or even harmful. In *The Fragile Species*, Lewis Thomas points out that treatments such as purges and leeches were finally abandoned only when they were objectively compared to simply allowing illness to take its course. Given time, we recover without any sort of intervention from most of the things that afflict us. Evolution has equipped our bodies with an elaborate array of natural defenses for dealing with injury or disease: bones knit, blood clots, antibodies seek out invading organisms, etc.

However, if someone who is purported to be a healer gives us herbal tea or sugar pills, or utters an incantation, or shakes a rattle over us, we are easily persuaded that the healing, when it comes, is the work of the healer. Depending on the culture, the healer will use props, such as a witch doctor's mask, or a stethoscope hung around the neck, to make the association more vivid. Once again, the belief engine is at work. The belief engine credits the treatment with any improvement that follows. It is the common logical fallacy *post hoc, ergo proper hoc*—"after it, therefore because of it." It could be the treatment, but in many cases it's just our repair system finally getting around to doing its job.

There is another part to the story, however. Once we are convinced of the healing power of a doctor or a treatment, something very remarkable happens: a sham treatment induces *real* biological improvement. This is the placebo effect. Healers have relied on the placebo effect for thousands of years, but until recently, it was usually referred to as the "mysterious" placebo effect. Scientists, however, are beginning to understand the complex interaction of the brain and the endocrine system that gives rise to the placebo effect.

People seek out a doctor when they experience discomfort or when they believe that something about their body is not right. That is, they suffer pain and fear. The response of the brain to pain and fear, however, is not to mobilize the body's healing mechanisms but to prepare it to meet some external threat. It's an evolutionary adaptation that assigns the highest priority to preventing additional injury. Stress hormones released into the bloodstream increase respiration, blood pressure, and heart rate. These changes may actually impede recovery. The brain is preparing the body for action; recovery must wait.

The first objective of a good physician, therefore, is to relieve stress. That usually involves assuring patients that there is an effective treatment for their condition and that the prospects for recovery are excellent—if they will just follow the doctor's instructions. Since we recover from most of the things that afflict us, the brain learns to associate recovery with visits to the doctor. Most of us start to feel better before we even leave the doctor's office.

Even in the first half of the twentieth century, most medicine was based on the placebo effect. Before 1940 about the only medicines doctors had in their bags were laxatives, aspirin, and sugar pills. Studies have shown, in fact, that if the patient believes the sugar pills will relieve pain, they will be about 50 percent as effective as the aspirin. The mechanism, however, is presumably quite different. Pain is a signal to the brain that there is a problem and something needs to be done about it. It's induced by prostaglandins released by white blood cells at the site of inflammation. Aspirin blocks the production of prostagladins.

The placebo, on the other hand, works by fooling the brain into thinking the problem is being taken care of. Once the brain is persuaded that things are under control, it may turn the signal level down by releasing endorphins, opiate proteins found naturally in the brain. Rather than blocking the production of prostaglandins, the endorphins block their effect. As powerful as the placebo effect can be, it is extremely doubtful that placebos can cause hair to grow on bald heads or shrink tumors, as some have claimed, but there is no doubt that placebos can influence the perception of pain.

Whether a person responds to a placebo depends almost entirely on how well the doctor plays his or her part. All the medical props, from the stethoscope to the framed medical school diploma, and all the soothing assurances given to the patient can be wiped out by an unguarded frown or a slightly raised eyebrow as the doctor goes over the patient's lab report. A placebo is most likely to work, therefore, if the doctor genuinely believes it to be a cure and communicates that conviction to the patient. Not surprisingly, then, those who imagine they possess miraculous healing powers, or truly believe they have discovered some wondrous cure that everyone else has overlooked, tend to be particularly good at invoking the placebo response. And often, like Samuel Hahnemann, they attract large followings.

LESS IS BETTER

In September of 1996, an international conference was held in Frankfurt, Germany, to celebrate the two hundredth anniversary of the publication of the German physician Samuel Hahnemann's law of similars, *similia similibus curantur*, or "like cures like," which forms the basis of homeopathy. The German health minister, Horst Seehofer, was on hand to greet the conferees. "Although there have been many attempts to prove otherwise," he told the appreciative audience, "the success of homeopathy cannot be denied." In Germany, homeopathy was now officially sanctioned mainstream medicine.

According to Hahnemann's law of similars, substances that produce a certain set of symptoms in a healthy person can cure those symptoms in someone who is sick. Although there are similar notions in the writings of Paracelsus in the Middle Ages and in Chinese medicine dating back thousands of years, Hahnemann seems to have reached this conclusion independently while trying to understand how quinine relieves the symptoms of malaria. He tried a little quinine on himself and experienced chills and fever—classic symptoms of malaria. From this single experience, he made the enormous leap to a general principle of medicine. Hahnemann's similarity approach ran counter to the prevailing medical view at the time, which was to prescribe treatments that appeared to suppress the symptoms. Both the similarity and opposition approaches were pitifully simplistic concepts in an age when knowledge of human physiology was still very primitive.

Hahnemann spent much of his life testing natural substances to find out what symptoms they produced and then prescribing them for people who exhibited those symptoms. Although the anecdotal evidence on which he based his conclusions would not be taken seriously today, homeopathy as currently practiced still relies almost entirely on Hahnemann's listing of substances and their indications for use.

Natural substances, of course, are often acutely toxic. Troubled by the side effects that frequently accompanied his medications, Hahnemann experimented with dilution. As you would expect, he found that with increasing dilution the side effects could be reduced and eventually eliminated. More remarkable, he also found that the more he diluted the medicine, the more his patients seemed to benefit. He came to the astonishing conclusion that dilution increased the curative power of his medications. He declared this to be his second law, "the law of infinitesimals." Less is better.

Hahnemann used a process of sequential dilution to prepare his medications. He would dilute an extract of some "natural" herb or mineral, one part medicine to ten parts water, or 1:10, shake the solution, and then dilute it another factor of ten, resulting in a total dilution of 1:100. Repeating that a third time gives 1:1000, etc. Each sequential dilution would add another zero. He would repeat the procedure many times. Extreme dilutions are easily achieved by this method.

The dilution limit is reached when a single molecule of the medicine remains. Beyond that point, there is nothing left to dilute. In over-the-counter homeopathic remedies, for example, a dilution of 30X is fairly standard. The notation *30X* means the substance was diluted one part in ten and shaken, and then this was repeated sequentially thirty times. The final dilution would be one part medicine to 1,000,000,000,000,000,000,000,000,000,000 parts of water. That would be far beyond the dilution limit. To be precise, at a dilution of 30X you would have to drink 7,874 gallons of the solution to expect to get just one molecule of the medicine.

Compared to many homeopathic preparations, even 30X is concentrated. Oscillococcinum, the standard homeopathic remedy for flu, is derived from duck liver, but its widespread use in homeopathy poses little threat to the duck population—the standard dilution is an astounding 200C. The *C* means the extract is diluted one part per *hundred* and shaken, repeated sequentially two hundred times. That would result in a dilution of one molecule of the extract to every 10^{400} molecules of water—that is, 1 followed by 400 zeros. But there are only about 10^{80} (1 followed by 80 zeroes) atoms in the entire universe. A dilution of 200C would go far, far beyond the dilution limit of the entire visible universe!

Hahnemann was presumably unaware that he was exceeding the dilution limit in his preparations. That calculation requires a knowledge of Avogadro's number, an important physical constant that allows you to calculate the number of molecules in a given mass of a substance. Although he was contemporary with Avogadro, Hahnemann published his major work, *Organon der Rationellen Heilkunde*, in 1810, one year before Avogadro advanced his famous hypothesis. It would, in fact, be another half century before other physicists actually determined Avogadro's number.

It is not difficult to see why Hahnemann became a popular physician. At that time in Europe, as in the United States, physicians still treated patients with bleeding, purging, and frequent doses of mercury and other toxic substances. If Hahnemann's infinitely dilute nostrums did no good, at least they did no harm, allowing the patient's natural defenses to correct the problem. As Hahnemann's reputation grew, patients expected to be cured by him. Belief evoked the placebo effect and allowed their body's own repair mechanisms to function unimpaired by stress.

SMART WATER

Hahnemann, as we saw, had no way of knowing that his medications exceeded the dilution limit, but his modern-day followers know. Avogadro's number is memorized in freshman chemistry. Homeopathists have calculated the dilution limit, and they agree that not a single molecule of the herbal extract or mineral could remain in their medications. But they insist it doesn't matter; the water/alcohol mixture in which the substance is dissolved somehow "remembers" the substance even after it has long since been diluted away. The process of shaking the solution between successive dilutions is presumed to charge the entire volume of the liquid with the same memory.

Although homeopathists have been administering this sort of "unmedicine" for two centuries, most scientists first became aware of their extraordinary claims with the 1988 publication in the prestigious British journal *Nature* of a paper by Jacques Benveniste, a French homeopathist. Benveniste reported that an antibody solution continued to evoke a biological response even if it was diluted to 30X or beyond. To say the paper was controversial would be to seriously understate the reaction of the scientific community. Many scientists felt the editor of *Nature* should have rejected the paper outright.

Benveniste, after all, had turned scientific logic on its head. Much of experimental science consists of devising tests to ensure that an experimental outcome is not the result of some subtle artifact of the experiment or its interpretation. "Infinite dilution" is one such test. If the observed effect does not go away when the concentration is reduced to zero, it is clear proof that the effect has nothing to with the substance being tested. But Benveniste claimed it proved the antibody had left an imprint of some sort on the solvent.

The editor of *Nature,* John Maddox, agreed that Benveniste's conclusion had to be wrong, but he nevertheless published the paper in the interest of open scientific debate. Publication of a paper in a peer-reviewed scientific journal, after all, does not amount to certification that it is correct. The reviewer's task is to make sure that there are no obvious errors and that the author has properly addressed any conflict with previous work. There is no way for the peer reviewer to know if the author has faithfully reported the results of the measurements or whether the instruments employed were faulty.

Maddox did take the unusual step of urging other scientists to replicate the Benveniste experiment, and in due course a well-respected group at University College London reported in *Nature* that they had repeated the Benveniste experiment as precisely as possible and found that "no aspect of the data is consistent with previously published claims." It had been Maddox's hope that exposing Benveniste's claims to scientific scrutiny would silence the homeopathists.

Homeopathists, however, continue to cite Benveniste's paper as proof of the law of infinitesimals and to concoct vague theories to account for this amazing result. These theories usually speculate that "information" from the active substance is retained in some way by the water. No one, however, has proposed an experiment to test these conjectures.

The reputed "memory" of water is only the first of a string of miracles that would be necessary for the law of infinitesimals to be valid. The over-the-counter homeopathic remedies available in health food stores, and increasingly in pharmacies, are generally in the form of pills. There are pills for everything from nervousness and swollen feet to menstrual cramps and flu. The pills are lactose tablets on which a single drop of the infinitely dilute solution has been placed. On the side of each bottle, a particular herbal extract or mineral, or some combination, is listed along with the dilution. Few people reading the ingredients realize that the pills do not actually contain any of these ingredients. The "solvent" is usually a water/alcohol mixture and quickly evaporates. Is the information that was somehow stored in the water now somehow transferred to the sugar? Does sugar remember the same way water remembers? And when the pill is swallowed, how is the information conveyed to the cells of the patient? And why do the medications only remember the good stuff and not the side effects? One miracle may reveal new science, but a string of miracles suggests a delusion. All the pills look alike—and in fact, they are all alike. Just sugar pills. There is no medicine in the medicine.

Many people still prefer their homeopathic medications the old-fashioned way, carrying out the sequential dilutions precisely as instructed by a homeopathist. Even the number of times the solution is shaken between successive dilutions is prescribed. Hahnemann initially shook, or "succussed," the solution ten times but later determined that four worked better. Preparing their own medications seems more natural to many patients, and the mystical ambiance generated by the ritualistic dilution procedures gives them a feeling of being in control.

The most authoritative reference for homeopathic home treatment is *Healing with Homeopathy* by Jonas and Jacobs. If you look up diaper rash, the authors recommend keeping the affected area clean and dry and prescribe *rhus toxicodendron*, better known as poison ivy. That, of course, is just an application of Hahnemann's law of similars. Baby's got a rash? Treat baby with something that causes a rash. It's fortunate for baby that Hahnemann also discovered the law of infinitesimals. The recommended dilution is 30C, or thirty sequential dilutions at one part in a *hundred*. That would result in one part medicine to one-followed-by-sixty-zeroes parts of the solution. That's more molecules than there are in the entire solar system. No matter. If baby is kept clean and dry, the rash will heal itself.

For children's diarrhea, the recommended medication is arsenic trioxide, which also finds use as a herbicide and as rat poison. No cause for concern. The recommended dilution once again is 30C. Alternatively, a vastly more concentrated solution of 12X can be used, but it's necessary to compensate for the higher concentration by administering it twice as often!

The standard homeopathy joke concerns the patient who died of an overdose after taking ordinary water by mistake. If this leaves you feeling dizzy, the recommended treatment for dizziness is crude oil—at a dilution of 30C.

Benveniste has since gone much further. He now claims to have discovered that the information is stored in the water in the form of electromagnetic waves that can be picked up by a coil surrounding the water. The information, according to Benveniste, can be stored in a computer and transmitted over the Internet to activate water anywhere in the world.

This is the point at which everyone is supposed to realize how ridiculous this is and share a good laugh. But homeopathists don't laugh. As preposterous as these claims sound to most scientists, they are taken quite seriously by some. Wayne Jonas, the nation's leading homeopathist and director of the National Institutes of Health (NIH) Center of Alternative Medicine, writing in *Nature Medicine*, solemnly described the infinite dilution claims of Benveniste as

> one of the deepest and most difficult enigmas in modern biology and medicine. Is it possible that specific non-molecular information can be stored and transmitted through water or wires as claimed in homeopathy? Even though this concept is implausible, the potential implications it holds for understanding basic biological and cellular communications are enormous.

That much at least is true. If the infinite-dilution concept held up, it would force a reexamination of the very foundations of science.

Meanwhile, there is no credible evidence that homeopathic remedies have any effect beyond that of a placebo. If the FTC is willing to move against a company like Rose Creek Health Products that markets ordinary salt water as a dietary supplement, why shouldn't the same action be taken against companies marketing homeopathic solutions? In both cases extravagant claims of benefits are made for ordinary water that has not been altered in any way, and the market for homeopathic remedies is enormous compared to that of "Vitamin O."

Homeopathy, it seems, retains a special legal status: In 1938 U.S. Senator Royal Copeland of New York, a homeopath before he became a senator, slipped a provision into the federal Food, Drug, and Cosmetics Act granting homeopathic remedies a special exemption from FDA oversight. Unlike drugs, homeopathic remedies could be marketed without any proof of safety or effectiveness. This exemption, which lacks any rational justification, remains the law more than sixty years later.

The consequences can be thoroughly bizarre: SmithKline, the maker of Nicorette, a smoking-cessation chewing gum, was compelled to establish its effectiveness in extensive clinical trials before it could be marketed. On the same shelf in your neighborhood drugstore is another smoking-cessation gum called CigArrest, which underwent no clinical trials and for which there is no evidence whatever of effectiveness. Why was CigArrest spared from clinical trials? Because its manufacturers claim it's homeopathic. Does chewing gum, like water, have a memory? Alas, there is no test that can be applied to distinguish homeopathic chewing gum from flavored chicle.

This raises a fascinating question. Since homeopathic remedies are "infinitely dilute," how can it be proven that they are in fact homeopathic? Even if you believe that water can have a memory, unless the mechanism by which that memory is stored is known, it is an invitation to fraud. Indeed, if a government agency was required to certify to the truthfulness of homeopathic labeling, what would it test for? It would be like trying to prove that holy water has been blessed.

MAGNETIC ATTRACTION

Behind the current popularity of "natural" medicine is the notion that industrialization has left humans starved for natural energy that was once supplied by Mother Earth. Dietary supplements are promoted as replacing something that is absent or diminished in the modern urban environment. The "Vitamin O" ads, for example, warned that the concentration of oxygen in the atmosphere, particularly in cities, is seriously diminished and needs to be supplemented.

The same sort of concern underlies the extraordinary magnet fad. The buildings in an urban area are said to block Earth's magnetic field, causing "magnetic deficiency syndrome." The symptoms are the usual vague complaints of tiredness and inability to sleep or concentrate. Wearing magnets is described as a restorative, like taking the vitamins we fail to get naturally from processed foods. In fact, an urban environment has little effect on exposure to magnetic fields, which are extremely difficult to eliminate. But as with dietary supplements, the claim is also made that even stronger fields can assist in the healing of injuries.

On the "Health Report" segment of *ABC World News Tonight* for August 11, 1997, news anchor Forrest Sawyer described "bio-magnetic therapy" as "the latest fad in sports medicine. It's been called a secret weapon to deal with aches and pains. The only problem is separating the extravagant claims from medical reality." It's a problem that would not be addressed on this program.

The reporter is Juju Chang. "It's a multimillion-dollar business," she explains. "The magnets, which cost up to eighty-nine dollars each, are available in more than a thousand golf pro shops." The viewers are shown images of celebrity sports stars in action: Miami Dolphins quarterback Dan Marino, who used magnets to heal his fractured ankle, is throwing a touchdown; Yankee center fielder Bernie Williams, who has magnets strapped to his pulled hamstring, makes a leaping catch at the fence; and finally, pro golfer Bob Murphy tells the viewers he has more than a dozen magnets strapped to his aging body.

Chang says there are many explanations for how magnets work. That's the cue for the talking heads. First there is a plastic surgeon named Daniel Man who explains, "Magnets provide a static or magnetic force that allows changes in the tissue." Force on what? What changes? His head is replaced by that of physical therapist Raymond Cralle, who informs us that "magnets are another form of electric energy that we now think has a powerful effect on bodies." That didn't clear anything up either. There is yet a third talking head. This one belongs to none other than Bill Roper, CEO of the company that's selling those refrigerator magnets for eighty-nine dollars. "All humans are magnetic," he explains; "every cell has a positive and a negative side to it." This will be news to cell biologists.

Juju Chang's job, presumably, is to explain what's going on in language the audience can relate to. With the aid of animated graphics, she explains that "magnets create a weak electric charge that increases blood flow to the injured areas." The graphic shows a bunch of flashing arrows, representing the magnetic field, going into the painful shoulder of a cartoon figure. The cartoon shoulder turns red, showing that blood is attracted by the magnetic field.

The idea that blood will be attacted by a magnet because it contains iron is a common misconception. The iron in hemoglobin molecules is in a chemical state that is not ferromagnetic. Blood, in fact, is diamagnetic, which means that it is actually weakly *repelled* by a magnetic field. This is easy enough to check. An excess of blood in one area of the body causes a flushing or reddening of the skin. That's why the skin reddens when you apply heat; blood is diverted to the heated area to serve as a coolant. But placing a magnet against your skin produces no reddening at all.

So far the count is three celebrity endorsements, three "explanations" by people who are neither scientists nor medical researchers, and one cartoon of a magnet healing an injury by a process that is clearly wrong. Surely now, I thought, it's time for the token skeptic. Coming into the view of the camera is a man in a white lab coat. Is this a physician, or is it a scientist? Neither, it turns out. According to Chang, he's a physical therapist at Lenox Hill Hospital in New York who has done research on magnet therapy. He explains that he could find no physiological effects from the use of magnets. So what's his conclusion? "We need research to find out if magnets work." I thought that's what he was doing.

The camera turns back to Chang for her summation. "The effectiveness of magnets," she concluded, "has not been definitively proven." In the meantime, she cautions, "don't use magnets around credit cards or pregnant women."

Certainly, magnets can play havoc with credit cards, but I had never heard of magnetic fields being dangerous for pregnant women or anyone else. My guess was that the field of these magnets would not extend through the skin of a pregnant woman, or through a leather wallet to the credit cards for that matter. Magnetic fields are not blocked by leather or skin, but refrigerator magnets are deliberately designed to limit the range of the field, precisely to avoid problems like erasing the coded information on a credit card. Refrigerator magnets consist of narrow strips of alternating north and south poles. Right at its surface, the field of such a magnet may be thousands of times as strong as the Earth's magnetic field. But a very short distance away, depending on the width of the strips, the north and south poles will cancel. I suspected that therapeutic magnets were made the same way; if so, they would have a very short range.

Golf pro shops are a little pricey for me, but the largest department store in Washington, D.C., was offering a Thera:P magnetic therapy kit for only $39.95. The box said the magnets were 800 gauss. If true, that's sixteen hundred times greater than the Earth's field—but the field of a magnet is measured at its surface. The magnets were in little Velcro pouches that were to be attached to the injured area with velvet straps. They must look very dashing in the fitness center. I removed a couple of the magnets from their pouch and slid one across the other. Just as I suspected, I could feel them click into place each time the poles lined up. They were just like refrigerator magnets: narrow strips of alternating north and south poles.

To get an idea of how quickly the field dropped off, I stuck one of the magnets on a steel file cabinet. I then held sheets of paper between magnet and file cabinet until the magnet could no longer support itself. Ten sheets! That's just one millimeter! Credit cards and pregnant women are safe! The field of these magnets would hardly extend through the skin, much less penetrate into muscles. Not that it would make any difference if it did penetrate. In fact, just the thickness of the velvet straps to which the magnet pouches were attached reduced the strength of the field to the point that it would not pick up a paper clip. Even a pair of scissors on my desk had enough residual magnetism to do that. Not only are magnetic fields of no value in healing, you might characterize these as "homeopathic" magnetic fields.

No amount of paid advertising could have served the purveyors of therapeutic magnets as well as this ABC News "Health Report." Not just the celebrity endorsements, but the credibility of a popular and trusted reporter, backed by a major network news department, was worth millions. And yet there was no indication that ABC News had bothered to check the facts with anyone who understood either magnetism or biology. The story relied entirely on information supplied by the magnet therapists themselves. It was not news; it was advertising disguised as news.

The use of magnets to treat injury or illness seems to have periodic revivals. Natural lodestone, or "magnetite," was used by the famed alchemist and physician Paracelsus in the early sixteenth century. The mysterious action-at-a-distance suggested great power. The power, of course, was the power of the placebo. Magnetic cures were introduced into England a century later by Robert Fludd—the inventor of the perpetual motion gristmill we met in the first chapter. While his mill may not have worked, the placebo effect does—within limits. But Dr. Fludd declared that the magnet was a remedy for all disease, if properly applied. This apparently involved insuring that the patient was in a "boreal position," with the head north and the feet south, while being treated.

By far the best known of the "magnetizers" was Franz Mesmer, who carried the technique from Vienna to Paris in 1778 and became the rage of Parisian society. Dressed in colorful robes, he would seat patients in a circle around a vat filled with "magnetized" water. Magnetized iron rods protruding from the vat were held by the patients while Mesmer waved magnetic wands over them. Eventually, Mesmer discovered that it was just as effective if he left out the magnets and merely waved his hand. He called this "animal magnetism."

The parallel with homeopathy is fascinating. That which is thought by the healer to be the cure is eventually eliminated—with no reduction in effectiveness. Yet the belief of the healer is unshaken. Indeed, the healer becomes convinced that an even more powerful effect has been discovered.

Other Parisian physicians, of course, bitterly resented Mesmer, an outsider who was attracting their most affluent clients. Benjamin Franklin, the world's leading authority on electricity, who was in

Paris as a U.S. representative, suspected that Mesmer's patients did indeed benefit from the strange ritual because it kept them away from the bloodletting and purges of the other physicians. Eventually, at the urging of the medical establishment, Louis XVI appointed a royal commission to investigate Mesmer's claims. The commission included Franklin and Antoine Lavoisier, the founder of modern chemistry. The commissioners designed a series of ingenious tests in which some subjects were deceived into thinking they were receiving Mesmer's treatment when they were not, while others received the treatment but were led to believe they had not. The results established beyond any doubt that the effects of Mesmerism were due solely to the power of suggestion. The commission report, drafted by M. Bailly, an illustrious historian of astronomy, has never been surpassed for clarity and reason. It destroyed Mesmer's reputation in France, and he retired to Austria.

How remarkable that two hundred years later, with all that has been learned about both magnetism and physiology, magnetizers should still be able to attract a following. Alas, there is no government commission of Franklins and Lavoisiers to challenge their claims. On the contrary, magnetic therapy is just one of the huge and growing list of untested and unregulated "alternative therapies" that were given official recognition by Congress in 1992. A single paragraph in the report language accompanying the NIH appropriation bill that year created an NIH Office of Alternative Medicine. What is "alternative medicine," and what is it doing in NIH?

THE ALTERNATIVES

The daily calendar of events in the *Washington Times* for March 2, 1995, listed a morning press conference in the Dirksen Senate Office Building to release an NIH report on alternative medicine. Holding the press conference in the Dirksen Building meant a U.S. senator was sponsoring it. But why, I wondered, would NIH choose to release a report on Capitol Hill? Recalling the controversy in *Nature* over Benveniste's infinite-dilution claims, I wondered if that was the sort of thing the report dealt with. Perhaps there were other issues that would interest a physicist.

I arrived at the press conference early and signed in as press. I have learned that by representing myself as press, I am assured of getting copies of all the printed material. Since college professors dress no better than reporters, I am rarely challenged. There was no television coverage, and looking down the list of real reporters that had signed in ahead of me, I noticed that there were no major news organizations represented. The reporters were mostly from health magazines I'd never heard of and publications with odd New Age–sounding names. I began to think I'd made a mistake in coming.

I was given a copy of *Alternative Medicine: Expanding Medical Horizons*; it was the report of a workshop conducted under the auspices of the National Institutes of Health to lay out a proposed research agenda for the new NIH Office of Alternative Medicine. It was about the size of the District of Columbia phone book. I took a seat near the door so I could slip out if this turned out to be really boring, and in the time before the press conference, I leafed through the report. It covered everything from herbal medicine and acupuncture to faith healing and Lakota medicine wheels—a wide spectrum of therapies related only by the fact that they lay far outside the field of accepted medical practice. Although many of these practices were linked to very ancient healing traditions, they had been dressed up in language about energy fields and quantum uncertainty and relativity, and in many cases fitted out with new names. "Biofield therapeutics," for example, referred to the ancient practice of "qigong" or "laying on of hands." I looked up Mesmer in the index. It directed me to a section that explained that "animal magnetism" was another term for the "biofield," known to ancient Chinese doctors as "qi." There was no mention of Mesmer's unmasking by the royal commission. My quick survey of the report failed to turn up a hint of skepticism about any therapy.

There was no one at the press conference representing the NIH director, which seemed strange since this was an NIH report, but several of the report's authors were on hand. The host for the press conference turned out to be Senator Tom Harkin, the populist Democrat from Iowa. Harkin was introduced as the father of legislation mandating the creation of an Office of Alternative Medicine (OAM) at the NIH in 1992. He opened the press conference by praising the report, even while complaining about how long it had taken to complete it. The OAM, he groused, had gotten off to a slow start "due to opposition from traditional medicine." With the publication of this report, he predicted, the OAM would make rapid progress.

Harkin, I later learned, had been introduced to alternative medicine by fellow Iowan Berkeley Bedell, a former congressman who resigned from the House after he came down with Lyme disease. Bedell now claims to have been completely cured by eating special whey from Lyme-infected cows. Harkin, in turn, claims to have been cured of his allergies by swallowing vast numbers of bee pollen capsules. Ironically, the substance that launched Harkin on his alternative medicine crusade seemed to be one of the few alternative treatments left out of the OAM report. Bee pollen, it turned out, was under a legal cloud. The man who sold the bee pollen to Harkin was forced to pay a $200,000 settlement under a consent agreement with the Federal Trade Commission, for making false claims in an infomercial. The infomercial claimed, among other things, that "the risen Jesus Christ, when he came back to Earth, consumed bee pollen." Testimonials are usually allowed, but the FTC apparently concluded that a testimonial from Jesus Christ was going too far.

Perhaps the strangest part of the press conference consisted of brief statements by individual members of the editorial review board of what they saw as the most important issues for the Office of Alternative Medicine. One insisted that the number-one health problem in the United States is magnesium deficiency; another was convinced that the expanded use of acupuncture could revolutionize medicine; and so it went around the table, with each touting his or her preferred therapy. But there was no sense of conflict or rivalry. As each spoke, the others would nod in agreement. The purpose of the OAM, I began to realize, was to demonstrate that these disparate therapies all work. It was my first glimpse of what it is that holds alternative medicine together: there is no internal dissent in a community that feels itself besieged from the outside. It is the same bond that unites the cold fusion community. But what brings the practitioners of alternative medicine together as a community in the first place?

Alternative medicine covers a broad spectrum of seemingly un-related treatments ranging from the plausible to the preposterous. The preposterous therapies are those that cannot possibly have any physiological consequences. In homeopathy, there is no medicine in the medicine. In magnet therapy, as we have seen, there may be almost no magnetic field. At the plausible end of the spectrum are dietary treatments and herbal therapy. Ironically, however, therapies at the preposterous end of the spectrum are perfectly safe since they have no effect at all, while those at the plausible end can have serious consequences.

Scientific pharmacology, after all, emerged from the traditional empiricism of the herbalist. With the exception of antibiotics, vir-tually all medicinals are originally derived from angiosperms, or flowering plants. Secondary chemicals, mostly in the leaves and bark, are produced for protection. Many of these natural sub-stances, including some that in small doses serve as medicinals, can be highly toxic.

The herb ephedra, or ma huang, for example, which has been used as a medicine in China for centuries, has become increasingly popular in the United States as a weight loss aid and as a "legal high." It contains ephedrine, a stimulant that mimics the effect of hormones such as adrenaline and has a number of medical appli-cations. But ephedrine is the basis of the street drug Ecstasy, and ma huang has been marketed, with no attempt at subtlety, as "herbal Ecstasy." There have been more than eight hundred reports of adverse reactions to ma huang, including liver damage and stroke, and at least seventeen deaths. Adding to the risk is the variability of herbal products, which makes it difficult to control the dose.

Scientific pharmacology seeks to identify and purify the active agent in an herbal remedy. Only then is it possible to carry out controlled studies of its effects and dosage. Aspirin, for example, was originally found in the bark of willow. Not until it was isolated and synthesized a hundred years ago, however, did it become what it remains today—one of the safest, cheapest, and most effec-tive drugs known. It was a "wonder drug" before the term was invented.

Between homeopathy and herbal therapy lies a bewildering array of untested and unregulated treatments, all labeled *alternative* by their proponents. *Alternative* seems to define a culture rather than a field of medicine—a culture that is not scientifically demanding. It is a culture in which ancient traditions are given more weight than biological science, and anecdotes are preferred over clinical trials. Alternative therapies steadfastly resist change, often for centuries or even millennia, unaffected by scientific advances in the understanding of physiology or disease. Incredible explanations invoking modern physics are sometimes offered for how alternative therapies might work, but there seems to be little interest in testing these speculations scientifically. In the final chapter, we will examine the question of whether the "strangeness" of modern physics offers any support for alternative medicine.

"Natural" remedies are presumed by their proponents to be somehow both safer and more powerful than science-based medicine. Fortunately, most natural medicine is in itself relatively harmless, aside from the financial damage done by paying eighty-nine dollars for a refrigerator magnet or twenty dollars for a vial of salt water. By keeping people from seeking unneeded antibiotics or overdosing on cold pills, something like homeopathy may actually promote health among the not-very-sick-to-begin-with. It can, however, become dangerous if it leads people to forego needed medical treatment. Worse, alternative medicine reinforces a sort of upside-down view of how the world works, leaving people vulnerable to predatory quacks. It's like trying to find your way around San Francisco with a map of New York. That could be dangerous for someone who is really sick—or really lost.

The feelings of antiscience and technophobia that find expression in "natural" medicine have their antithesis in the worship of technology. While in this chapter we saw that many people seek only that which is natural, in the next chapter we will meet others who dream of life on an artificial world. It should not be surprising that those who love technology too much also succumb to voodoo science.

FOUR
THE VIRTUAL ASTRONAUT
In Which People Dream of Artificial Worlds

DREAMS OF A STATION IN SPACE

I RELUCTANTLY TOOK MY PLACE at the witness table before the House Subcommittee on Space and Aeronautics on April 9, 1997, to testify about the International Space Station. I would be about as popular at this hearing as a skunk that wandered into a garden party. Seated between the head of NASA's human space-flight program on my left and a former astronaut on my right, I had been invited as the token critic. I had testified on the space station before congressional committees many times in the past, but there had always been some hope that Congress might be persuaded to cancel the project. But in 1997, although the space station was years behind schedule and several times over budget, its support in Congress and particularly in the Space Subcommittee was stronger than ever. With the much-delayed launch of the first module now only months away, the question that concerned the Space Subcommittee was not whether the space station should be built but whether we should depend on the Russians, who were already many months behind in their commitments, or go it alone.

What most members of the Space Subcommittee wanted to hear was that the space station is so important to science that it should be built with or without Russian help. They would not hear that from me, or any other scientist I knew. The space station cannot be justified on scientific grounds. The science planned for the space station, however, is quite unlike the voodoo science we've discussed previously; it is not so much wrong as simply unimportant. The story of the space station is a story of misplaced dreams and unwarranted hype. It's voodoo science by press release.

In the early days of the space program, a permanently manned space station had seemed like an inevitable step in the conquest of space. From such a platform it would be possible to make astronomical observations free of distortion by Earth's atmosphere, monitor weather systems, relay communications around the globe, provide navigational assistance to ships and aircraft, and detect clandestine military operations. By 1984, however, these functions were all being performed by unmanned satellites. What's more, the robots were doing them far better and far more cheaply than would ever have been possible with humans.

A manned space station is simply not stable enough to obtain high-resolution images of either the heavens or the Earth; the slightest movements of the crew cause the center of gravity of the station to shift, and the rotating machinery necessary for life support produces vibrations that blur images under high magnification. A human crew must be resupplied and rotated as the stress of survival in the harsh environment of space saps their strength and resolve, while a robot satellite can be left unattended for years, powered only by sunlight. The dream of a space station, however, was too deeply embedded in the national imagination to go away. Each new administration would concoct a new justification.

The space station of Ronald Reagan's dreams was to be a sort of microgravity research and development laboratory that would lead the way to space manufacturing. "We can follow our dreams to distant stars," President Reagan told the nation in his 1984 State of the Union Address, "living and working in space for peaceful economic and scientific gain. Tonight, I am directing NASA to develop a permanently manned space station and do it within a decade." But the space station wasn't headed for the stars; it was aimed at low-Earth orbit, a region of space that has been so thoroughly explored that it is dangerously littered with debris left behind by hundreds of previous missions.

Space, they were fond of saying in the Reagan White House, is just another place to do business. There were claims that in microgravity it would be possible to manufacture more perfect ball bearings, develop new alloys, grow more perfect semiconductor crystals, and create new drugs. The same claims had all been made a decade earlier to justify the space shuttle. Alas, the prospects had been wildly exaggerated by space enthusiasts. The force of gravity is just too weak compared to the electromagnetic forces that bind atoms together to have any significant effect on manufacturing processes. And what could be manufactured that would justify the transportation costs? On the market today, gold closed at $311 per ounce. The cost of launching that ounce of gold into low-Earth orbit using the shuttle would be about $830—and it would cost about the same to bring it back. If there was gold for the taking in low-Earth orbit, it would not pay to go get it.

Just eighteen months after President Reagan's space station speech, the Soviet Union launched Space Station Mir. As a critical vote on the space station was being debated in Congress, a television commercial by the aerospace giant McDonnell-Douglas began running several times a day on all the major networks. It was thirty seconds of television-commercial art at its most powerful. The stars are seen against the blackness of space. The edge of a large space station rumbles into view, blotting out stars as if passing just over your head. With the muffled hum of machinery in the background, a narrator's voice solemnly intones, "Right now, miles above the Earth in a manned space station, experiments are being conducted that could cure major diseases, new and valuable alloys are being created, and new scientific data that could literally change the course of history are being collected every minute." There is a pause just as a large red star on the side of the passing space station comes into view. "Shouldn't we be there too?" the voice asks. As the scene fades, the sound of voices speaking in Russian seem to come from the space station—followed by laughter.

Today, more than a dozen years later, there are still no cures, no new alloys; indeed, no field of science or technology has been affected in any significant way by microgravity research either on Mir or the American space shuttle. I once asked the former head of the Soviet space science program what cosmonauts on board Mir find to do all day. "They try to stay alive," he replied. On Mir, that hasn't always been easy.

Faced with these hard realities, the business community showed little interest in the space station plan. Reagan's vision of manufacturing in space faded, while cost estimates for the station kept growing. The original estimate had been $8 billion, but more than $12 billion had already been spent on design alone with no hardware in sight. The space station seemed to have become nothing more than a make-work welfare program for the aerospace industry, and a move in Congress to kill it failed by a single vote. It was clear by the time George Bush entered the White House in 1989 that the program would be killed unless a new justification could be found.

President Bush offered a more romantic dream. In a speech delivered from the steps of the National Air and Space Museum in Washington on the twentieth anniversary of the *Apollo 11* Moon landing, Bush called on the youth of America to "raise your eyes to the heavens and join us in a great dream—an American dream—a dream without end." Summoning up the voyage of Christopher Columbus to the New World, he called for a return to the Moon and a manned expedition to Mars. "Like Columbus," he said, "we dream of shores we've not yet seen."

President Bush had found a new mission for the space station: it would be used to prepare for an expedition to Mars. It would become a space-medicine laboratory dedicated to finding ways to prevent the debilitating effects of prolonged exposure to a space environment. It had taken the Apollo astronauts three days to reach the Moon. Mars would take nine months; the round trip, almost three years. Unless ways could be found to deal with the physiological consequences of extended stays in space, it was not at all clear that an astronaut who reached the red planet would be in any condition to go exploring. The Soviets were already learning how difficult extended missions in space could be. Courageous cosmonauts were testing their endurance on Space Station Mir. Nine months on Mir left them too weakened to stand on their return to Earth. Recovery was slow and perhaps incomplete.

By then known as Space Station Freedom, the station was completely redesigned to match its new mission. It was also downsized, reducing the crew from eight to four, to compensate for the still-escalating cost estimates. Meanwhile, NASA began preliminary planning for a manned mission to Mars. Midway through the Bush administration, however, the cold war came to an abrupt and unexpected end with the collapse of the Soviet Union.

The space station was in trouble again. The end of the cold war meant an end to the competition that had driven the space program for thirty-five years. To add to the space station's woes, cost estimates had continued to spiral upward in spite of the downsizing. Moreover, initial cost estimates for the mission to Mars ranged from $500 billion to $1 trillion, an unthinkable sum in the absence of a compelling need. Plans for an astronaut expedition to Mars were dropped. The high cost of transporting things into space threatened to end the space station program.

When Bill Clinton became president in 1993, the space station was again drifting without a mission. Dan Goldin, the maverick administrator George Bush had put in charge of NASA, was the only high-level Bush appointee left in place by the Clinton administration. The rumor in Washington was that Goldin would be kept long enough to kill the space station. Clinton personally had no interest in the space program; he had been elected to revive the sagging economy. But the space station, although it still existed only on paper, had by now become part of the economy, employing thousands of people in a depressed aerospace industry and pumping money into almost every congressional district in the nation. It was now Bill Clinton's turn to invent a new mission for the space station. His decision would not be driven by scientific concerns.

World peace was chosen as the new theme of the newly renamed International Space Station (ISS). President Clinton offered it as a model for a new era of international cooperation on major science projects. Fifteen other nations joined the United States as "partners" in the ISS, contributing various components of the giant Tinkertoy that would be assembled in space. But their participation was mostly symbolic; everyone knew it was an American project and America would continue to bear most of the cost. The Russians, who still maintained the aging and precarious Mir, would be brought in as a "full partner"—but with financial assistance from the United States The transition from Space Station Freedom to the International Space Station required yet another redesign, both to accommodate Russian capabilities and to again trim the relentlessly escalating cost. The unstated objective of the Clinton administration was to provide employment for Russian space experts who might otherwise be tempted to sell their services to rogue nations seeking missile technology.

That's where things stood when I began my testimony. In these hearings, witnesses are usually given only a few minutes to summarize lengthy written testimony. I explained that the enormous benefits from the space program, both in satellite technology and basic science, were unrelated to human space flight. No field of science, I told them, had been significantly affected by the microgravity research carried out on the shuttle or on Mir. The space station, I ended, stands as the greatest single obstacle to the further exploration of space. There was much more that I wanted to tell them; I wanted to explain why the era of human space exploration had ended twenty-five years earlier and was unlikely to ever come back. But to explain that, I would have had to go back forty years to the launch of *Sputnik*.

BEYOND THE IONOSPHERE

We knew almost nothing in 1957 about what conditions are like only a few hundred miles above the surface of Earth, much less in the rest of the solar system. Our knowledge of other planets was limited to the images seen through Earth-based telescopes. It was still widely believed that there were seasonal changes on Mars related to some sort of plant life, and some scientists clung to the notion that there might be civilization on the red planet. Venus, perpetually shrouded in clouds, was often pictured as a planet of fog and swamps.

Sputnik I, launched into orbit on October 4, 1957, carried no scientific instruments, but its beeping as it passed overhead every ninety minutes could be picked up on thousands of short-wave radio sets. It was a terrible shock to Americans, many of whom refused to believe the Soviets could have such sophisticated technology. Just one month later, *Sputnik II* was sent into orbit. It was apparent that *Sputnik II* was sending back scientific data—the first information to reach Earth from an instrument beyond the ionosphere. The official Soviet announcements gave little hint about what scientific instruments were on board.

Herb Friedman of the Naval Research Laboratory, one of the pioneers of radio astronomy, was on his way to a space science conference in Leningrad when *Sputnik II* was launched. When he arrived, he asked his Russian hosts if they could tell him what instruments were on board. They did better than that. They took him to a trade fair on the outskirts of the city where a cutaway replica of *Sputnik II* was on display. He was allowed to take a closer look: a tape recorder, a radio transmitter, and two glass Geiger tubes. He wrote down the tube number. The following day, he was walking down the street near his hotel with John Simpson, another American space science pioneer. They passed a store that specialized in demonstration equipment for schools. In the window was a Geiger tube identical to the ones in the *Sputnik II* replica. They went in and asked if it was possible to buy such tubes. The clerk checked and found they had them in stock. Herb bought two, the clerk wrapped them in a copy of *Pravda*, and Herb slipped them into his coat pocket. They were still there when he flew back to the United States.

Shortly after his return, he was visited by someone from naval intelligence, who was interested in anything Herb might have heard about what kind of scientific data *Sputnik II* was sending back. So far, he said, the CIA had learned nothing. Herb reached into his coat pocket and took out the two Geiger tubes, still wrapped in a copy of *Pravda*. The astonished agent asked if he could borrow them. Several months later, Herb had just about forgotten the whole episode when an armed security courier showed up at his office to deliver a package marked TOP SECRET. The two Geiger tubes he had picked up in a shop in Leningrad for one ruble had been returned.

The following June, Friedman and other American space scientists were back in the Soviet Union for another conference, but this time it was the Americans' turn. On January 31, 1958, just four months after the launch of *Sputnik I*, the U.S. Army had launched the *Explorer I* satellite, which carried its own Geiger counters. The high point of the conference was James Van Allen, a physics professor at the University of Iowa, announcing the discovery of the first of the two "Van Allen radiation belts," doughnut-shaped bands of charged particles trapped above the equator by Earth's magnetic field. It was the first major scientific discovery in the exploration of space. The Soviet leader of the Sputnik program was visibly crestfallen. The cold war was fought with symbols, and *Sputnik* had given the Soviets an enormous propaganda victory. Scientifically, however, the United States had already taken over the lead; it was never again relinquished.

But why had *Sputnik II* failed to detect the Van Allen belts? The Geiger tubes it carried should have recorded the radiation. Therein lies another lesson. Just prior to launch, technicians had discovered a problem with the tape recorder. Without it, only data gathered while the spacecraft was in line-of-sight from the receiving station could be retrieved. The scientists requested permission from Soviet premier Nikita Khrushchev to postpone the launch until it could be fixed. Khrushchev refused; it was the eve of an important meeting of heads of state, and he wanted to announce a successful launch. If the Soviets had been able to collect data from a complete orbit, they would have discovered the Van Allen radiation belts. At the very dawn of the space age, politics was already getting in the way of scientific discovery.

On March 26, 1958, not quite six months after *Sputnik I*, President Dwight Eisenhower shared with the American people a remarkable document prepared by his Science Advisory Committee, chaired by James R. Killian Jr., the president of MIT. Written in clear, nontechnical language, *Introduction to Outer Space* explained in simple lay terms "why satellites stay up," using the example of a stone thrown so hard that its arc carries it over the horizon, missing the Earth. The satellite is perpetually falling.

A rocket, the report explained, gets its thrust by exhausting material backward. To truly break its bonds to Earth, the rocket must accelerate to an escape velocity of about 25,000 mph. It will continue to be slowed down by Earth's pull, but with gravity getting weaker and weaker, it will never come to a stop and fall back. "Although the basic laws governing satellites and space flight have been well known to scientists ever since Newton," the report said, "they may seem a little puzzling and unreal to many of us. Our children, however, will understand them quite well."

The report suggested that someday it would be possible to send a manned expedition to the Moon but stopped short of recommending similar expeditions to other planets. A section of the report titled "A Message from Mars" captured the feelings of most scientists: "This all leads up to an important point about space exploration. The cost of transporting men and material through space will be extremely high, but the cost and difficulty of sending *information* through space will be comparatively low." The exploration of the solar system, the report predicted, would be conducted with instruments that need not return. The information gathered by these instruments would be sent back to Earth as radio waves.

Another science advisory panel, this one headed by professor Jerome Wiesner of MIT, made the same point in a report to John F. Kennedy shortly before his presidential inauguration in 1961. Whatever needed to be done in space, the Wiesner report said, could be done more effectively and much less expensively with unmanned spacecraft. The clear logic of the Wiesner report was swept away just three months later when Soviet cosmonaut Yuri Gagarin was sent into orbit. The imagination of people everywhere in the world was captured by the image of a man who had, however briefly, broken the bonds to Earth. In a war of symbols, the American lead in space science meant little. The United States had been handed its second devastating defeat in space.

President Kennedy, whose gift was to understand the deeper aspirations of the people, recognized at once the powerful symbolism of human spaceflight. Ignoring the advice of the Wiesner panel, he immediately instructed NASA to shift the focus of the space program from instruments to men of flesh and bone. Within a month, he had committed the United States "to put a man on the Moon and return him safely to Earth before this decade is out." In the context of the cold war, it was a bold gamble. The United States was challenging the Soviet Union to a race to the Moon — and the Soviets had a head start.

The advance of knowledge was almost incidental, merely a prop to justify a display of national technological prowess. It is a role that would be perfected a decade later in the missions of the space shuttle; the most inconsequential research results would be promoted as dazzling accomplishments. The motivation would be political rather than scientific.

A year after President Kennedy laid down his challenge to the Soviets, John Glenn became the first American to orbit Earth. On his return, Glenn was given the sort of welcome that had greeted Lindbergh, including the ticker-tape parade up Broadway. Like Yuri Gagarin, the first Soviet cosmonaut to orbit Earth, John Glenn would not return to space — at least not for many years. Both were cast in the role of national heroes, too valuable as symbols to be exposed to further risk.

There was, of course, another astronaut who successfully rode a Mercury capsule into orbit before John Glenn — Ham the chimpanzee. Both Ham and Glenn would end up in Washington: Glenn in the U.S. Senate, Ham in the National Zoo. Ham died a short time later without ever returning to space.

Later that same year, with far less acclaim, an unmanned American probe, *Mariner 2,* flew by the cloud-shrouded planet Venus. It was the first successful planetary flyby by any nation; two earlier Soviet attempts had failed. *Mariner 2*'s instruments gave us our first direct information, other than the images of Earth-based telescopes, about conditions on another planet. Few people seemed aware of the significance of what was happening: it was our robots, not our astronauts, that were exploring the solar system. *Mariner 2* had traveled to Venus, a hundred million miles from Earth. John Glenn had been no further from Earth than New York is from Baltimore.

Seven years later, on July 16, 1969, President Kennedy's promise was fulfilled. Incredibly, Neil Armstrong stood on the Moon, 240,000 miles from Earth. How can *Apollo 11* be described? It was a feat of skill and daring unmatched in history. The Apollo moon landing transcended the struggle between the United States and the Soviet Union for world domination. It was a source of pride and inspiration for the whole human race, symbolizing the heights that humans are capable of reaching and overshadowing every space mission before or since. But it was also an incredible political triumph. The entire world stood in awe of the American achievement. The trauma of Sputnik was finally behind us.

Ironically, it was now the Soviets who had misjudged the symbolic importance of human presence in space as a measure of national power. Following the advice of their scientists, they had put their energies into robotic exploration. Just one day after the American landing of *Apollo 11*, an unmanned Soviet spacecraft, *Luna 15*, arrived at the Moon. Meant to return lunar soil samples, *Luna 15* crashed while attempting a lunar landing. A year later, however, *Luna 16* did return samples, and shortly thereafter *Luna 17* deployed a robot vehicle that traveled several kilometers over the lunar surface transmitting live pictures back to Earth. The Soviets continued their lunar exploration using robots until 1976, even returning with samples taken from seven feet below the surface. It was a stunning technical achievement, and the scientific returns may have surpassed those of the Americans, but few people today even remember it happened. It was clear that in the exploration of the Moon the world judged the United States to be the winner.

The *Apollo 11* Moon landing had been hailed as the opening of a new era of human exploration. There would be five more astronaut missions to reach the Moon, ending with *Apollo 17* in 1972. Lunar colonies, expeditions to Mars and beyond — it all seemed just around the corner. Could the stars be far behind? No one could have imagined that it was not a beginning but an end. Yet since that time, no human being has ventured beyond the relative safety of low-Earth orbit, and no such missions are scheduled. As if by mutual agreement, the contest between the United States and the Soviet Union for supremacy in space would now be played out on a field just beyond the ionosphere. The era of exploration of the solar system by human astronauts was over. It had lasted only three and a half years and never extended beyond the Earth's own moon.

THE RETREAT TO LOW-EARTH ORBIT

What had we learned that led to the quiet shelving of the dream of interplanetary travel by humans? Aside from our own moon, the only conceivable destination for human explorers is Mars. Even the most ardent champions of human exploration of space concede that for the forseeable future there is simply nowhere else to go. Conditions on the other planets or their moons are either too hot or too cold, or the gravity is too strong, or the radiation levels too intense, for humans to ever set foot on them. And even Mars is no Garden of Eden.

In 1957, the first year of the space age, it was apparent that the radiation hazard outside the Earth's protective magnetosphere was a serious concern. The continuous background level of solar and galactic radiation in interplanetary space far exceeds the radiation limits allowed for workers in the nuclear industry. Giant solar storms, which produce huge bursts of charged-particle radiation, are a potentially fatal threat; even unmanned communications satellites have been disabled by solar storms.

In the low-Earth orbits accessible to the space shuttle, only two or three hundred miles up, Earth's magnetic field provides some protection from charged-particle radiation, but long-duration interplanetary travel is another matter. On a three-day trip to the Moon, it's possible to cross your fingers and hope that there won't be a solar storm, one of which occurs about every six months, but a round trip to Mars would take two to three years.

Satellites in orbits near the sun could detect solar flares and relay a warning to the astronauts, giving them just minutes before the arrival of a burst of charged particles traveling a little slower than the radio waves. Time enough to hop into a "storm shelter"—a sort of lead-lined coffin—until the storm blows by. A sufficiently large solar storm would be fatal to an astronaut caught outside his shelter. But even if the astronauts avoid the occasional solar storm, they will still receive a heavy dose of background radiation, not only from the Sun but from the rest of the galaxy.

Galactic radiation (radiation coming from outside the solar system) includes extremely energetic heavy particles—the nuclei of heavy (high-Z) elements. A recent National Academy of Sciences report concludes that in the course of a round trip to Mars, every cell in the body would be traversed at least once by a high-Z, high-energy particle. Surprisingly little is known about the effects of high-Z radiation on living cells. High-Z radiation is quite unlike the nuclear radiation that concerns us on Earth, which consists of neutrons, electrons, alpha particles (the helium-4 nucleus), and gamma rays, which are very high-energy photons. There are few particle accelerators capable of producing high-Z radiation for animal studies, but there is every reason to expect the damage to cells to be massive. Cancer is a long-term threat, but damage to the central nervous system would be of more immediate concern. And it wouldn't be much better when the explorers got to Mars: half the galactic and solar radiation would be blocked by Mars itself, but as the *Mars Global Surveyor* confirmed in 1997, Mars has no magnetic field and almost no atmosphere to shield against charged-particle radiation. Astronaut explorers on Mars could not venture far from their shelter.

A more unexpected discovery is the severity of health effects from even a relatively short exposure to zero gravity. In the first heady days of the space age, there had been speculation that heart patients might someday be sent into orbit to rest their hearts, which would not need to pump blood against the force of gravity. On the contrary, forty years of studying of humans in a microgravity environment has turned up one harmful effect after another. Not only is the heart severely stressed in zero gravity, bones lose calcium, muscles atrophy, the immune system is depressed, diarrhea is endemic, sleep cycles are disrupted, and there are frequent bouts of depression and anxiety. Measures such as a rigorous exercise schedule slow the effects down but do not prevent them.

None of these problems prevents human space travel, but they add enormously to its dangers and its cost. Shielding could be added, increasing launch costs, and spacecraft could be designed to rotate like a huge wheel as Arthur C. Clarke had imagined, generating artificial gravity at its perimeter, again at great cost. In short, the space environment has turned out to be a far greater obstacle to human space travel than anyone realized prior to Sputnik—an unforeseen scientific discovery that calls for a reexamination of our priorities in space. It is a major reason why human activity in space is now confined to low-Earth orbit.

Nevertheless, for those who dream of human civilization expanding beyond the confines of Earth, Mars is the only realistic possiblity. It is also the most scientifically intriguing destination in the solar system. Will we, could we, should we go there? From a scientific standpoint, we already have.

THE MARTIAN CHRONICLES

The entire world visited Mars in the summer of 1997, taken there on the Fourth of July by a cocker spaniel-sized robot named Sojourner. The bleak, desiccated landscape we saw through Sojourner's eyes must once have been warm and wet; there was clear evidence that great torrents of water had long ago tumbled over the boulder-strewn surface. Had the rivers of Mars, like streams on Earth, teemed with living creatures? Could there be simple forms of life still surviving beneath the barren surface? Something in us longs to know.

With little atmosphere and no magnetic field to shield them, humans could not long endure the cosmic and solar radiation that sterilizes the Martian surface. Locked in space suits on that airless world, astronauts would have no sense of touch or smell; the only sound would be the faint, almost imperceptible, rumble of the thin Martian wind. They would have only the sense of sight, and Sojourner had better eyes than any human.

We are returning with more sophisticated robots. If life is found on Mars, it will be our first contact with living organisms that do not share a common genealogy. As surely as DNA analysis can establish that I am the father of my sons, it demonstrates and quantifies the kinship of all life on Earth; 98.4 percent of human DNA is identical to that of the chimpanzee (were it not for the other 1.6 percent, we would be obliged to offer chimpanzees a seat in the United Nations). Perhaps 50 percent of our genes are shared with lowly yeast. We have genes in common with bacteria, even though our branch of the family, the eukaryote, split off billions of years ago from the prokaryote branch that includes bacteria. No record remains of what went before, but science gives us no reason to invoke any powers beyond the laws of chemistry to explain the origin of life, nor any reason to doubt that the wondrous variety of complex life forms that inhabit Earth today arose by entirely natural causes: the mutation/selection cycles of Darwinian evolution. Extraterrestrial life, however simple, would give us insight into the course of a wholly different evolutionary experiment. How common is life in the universe? Given favorable conditions, is the appearance of life inevitable? How many ways could the experiment of life turn out?

The Copernican solar system, Darwinian evolution, galaxies, quantum mechanics, the big bang, and the genetic code were all intensely uncomfortable discoveries, diminishing the specialness of humankind and exposing cherished myths to ridicule. They were grudgingly accepted, not because they are pleasant but because they are true. The discovery of extraterrestrial life would force us, as every great scientific discovery has, to reexamine the way we think about the universe and our place in it.

A year after Sojourner's great adventure, more than seven hundred people from all walks of life and forty nations gathered in Boulder, Colorado, to found the Mars Society. They listened to four days of talks about Mars, but it was not only thoughts of Martian life that fired their imaginations. Their dream was to transplant terrestrial life to the red planet.

The organizer was a forty-six-year-old engineer named Robert Zubrin. Most of them knew the plan already; they had read Zubrin's book, *The Case for Mars*. Zubrin could make the Mars dream sound possible, even inevitable. "We need a central overriding purpose to drive our space program forward," he told the spellbound audience. "At this point in history, that focus can only be the human exploration and settlement of Mars." Zubrin has proposed "terraforming" Mars, creating a new Earthlike home. Hovering over Zubrin as he spoke was the ghost of Gerard K. O'Neill. Let me explain.

By 1972 the seemingly boundless technological optimism generated by the Apollo moon landings had given way to a tide of

Malthusian pessimism. A study commissioned by the Club of Rome, an international assembly of business leaders, resulted in the publication of *The Limits of Growth,* by Dennis Meadows and colleagues at MIT. Its somber message resonated with a generation turned off to technology by the Vietnam War. The industrialized nations, Meadows warned, were depleting Earth's resources and destroying the environment with consequences that would surely lead to disaster unless the nations of the world adopted policies of austerity and population control.

Technological optimists were horrified at what they regarded as negative thinking. Only with the prosperity that accompanies un-shackled industrialization, they believed, could the problems of the world be solved. Gerard O'Neill, a Princeton physicist, was convinced that limiting growth would inevitably lead to an authoritarian government presiding over the distribution of resources and regulating the right to reproduce.

There was, however, no denying the finiteness of Earth. The only way growth could be sustained indefinitely, O'Neill reasoned, was to expand off the planet. In his popular 1976 book, *The High Frontier*, O'Neill proposed the construction of "islands in space." He imagined them as gigantic hollow cylinders, rotated about their axis to create artificial gravity for people living on the inner surface. Within the limits of current technology, he calculated, these cylinders could be four miles in diameter and twenty miles long, with a total "land" area of five hundred square miles supporting a population of several million people. The residents of his colonies would be engaged in the construction of additional colonies from materials mined from the Moon and asteroids. By off-loading the excess population of Earth to his "islands," O'Neill imagined, we could have unlimited industrial growth without environment damage.

The media and the NASA public affairs office loved O'Neill's space colonies, and with NASA funding he supplied ever more detailed plans and drawings of life on "Island One." It always looked like an affluent suburb, with trees and lawns and lakes—but with no horizon. If you looked up, you would see those living halfway around the cylinder. The idea attracted a cultlike following of dedicated supporters called the L5 Society. They tirelessly roamed the halls of Congress lobbying for federal funding to make O'Neill's colonies a reality.

The name "L5 Society" was taken from a stable point on the orbit of the Moon equidistant from the Moon and Earth. Discovered by the French mathematician Lagrange, it was called the Lagrange 5-point, or simply L5. A satellite at the L5 point could remain in place indefinitely without the need to expend fuel to keep it from drifting. It would be the logical place to erect a colony. The L5 Society wasn't talking about some far distant future; its slogan was "L5 by '95." But how to justify such an enormous undertaking?

O'Neill, driven by ethical and religious principles, had simply argued that growth is inherently good, but even the starry-eyed members of the L5 Society knew that there had to be some economic justification. The expedition of Columbus, after all, was not funded out of idealism but by a desire to reap the riches of the East. To provide an economic rationale, the L5 Society tried to link the idea of space colonies to proposals for solar-power satellites. The citizens of Island One would be engaged in the construction and maintenance of these satellites. Gigantic arrays of solarvoltaic collectors would convert sunlight into electrical energy, which would somehow be beamed back to Earth, perhaps as microwaves. With solar arrays in space, there would be none of the earthly concerns about nighttime and cloudy days, or damaging hailstorms, or the difficulty of keeping acres of semiconductor surface free of dust. Beaming such huge amounts of energy back to Earth, however, assuming it could be done, raised serious safety concerns and questions about possible use as a weapon.

No matter. No one talks seriously about space colonies any longer. It was not that a space colony couldn't be built—it would violate no laws of physics—but the future must also conform to the laws of economics. The detailed calculations of feasibility cranked out by O'Neill's institute at Princeton had been based on NASA's wildly optimistic estimates of the cost of launching material into space using the space shuttle. These estimates had been used by NASA to persuade Congress that building the shuttle would reduce launch costs.

But it was not economics that drove the shuttle program. The adoration bestowed on the Apollo astronauts by a grateful public had persuaded space officials that humans must play a role in all future space missions. They were convinced that a public weaned on *Star Trek* would not support a space program that did not feature humans. All existing launch systems were to be replaced with reusable piloted vehicles. Even unmanned planetary probes would be relaunched from the shuttle once it was in orbit. Nothing would go into space without images of astronauts showing up on the evening news.

It was a costly miscalculation. The shuttle would never come close to meeting the nation's launch needs. The queue of postponed and canceled missions was already strangling the U.S. space effort in 1986 when the *Challenger* disaster brought a halt to launches for three long years. There was no backup. Like some mad general burning bridges behind his army to prevent retreat, the assembly lines for the great Saturn rockets that had delivered astronauts to the Moon had been demolished and the plans destroyed. Far from reducing launch costs, the shuttle turned out to be the most expensive space-launch system ever devised. O'Neill's own calculations, corrected for the actual cost of shuttle launches, revealed the idea of space colonies to be hopelessly unrealistic. Reality is the International Space Station—seven astronauts crammed into a can as spartan as Alcatraz, at a cost that threatens to bankrupt the space programs of sixteen nations. Island One it is not.

The L5 Society faded into oblivion. But Robert Zubrin has taken up the cause of establishing extraterrestrial colonies. Twenty years after publication of *The High Frontier*, Zubrin would write his own book, *The Case for Mars*, and start his own organization—the Mars Society. Members of the Mars Society focus on the dream. They feel their feet crunching into the sands of Mars, while the most daunting technical challenges are swept aside with simplistic solutions.

Human travel to Mars, according to Zubrin, would be easier than a trip to the Moon, and we could do it on a shoestring. The trick, he says, is to follow the example of Lewis and Clark and "live off the land." It evokes an image of explorers killing wild game and digging for roots, but what Zubrin is proposing is to manufacture methane rocket fuel on Mars for the return trip to Earth, using the thin CO_2 atmosphere as raw material. Manufacturing rocket fuel from CO_2, however, requires, among many other things, an enormous amount of energy. No problem; the expedition will take along a portable nuclear reactor.

Some idea of how difficult it would be to sustain an extraterrestrial colony on Mars was provided by Biosphere 2, a three-acre, sealed-off mini-world built in the Arizona desert (Biosphere 1 was taken to be Earth itself). The gleaming glass-and-steel enclosure rising above the Arizona desert was bankrolled by Texas billionaire Edward P. Bass, the New Age heir to the Bass oil fortune. Bass had fallen under the spell of a group that believed Earth was about to be consumed by civil unrest, whereupon the group members would escape to Mars.

Hailed by some in the media as a bold scientific experiment, four men and four women dressed in *Star Trek*–style uniforms waved to the cameras and marched into the human terrarium on September 26, 1991. The airtight door was sealed behind them. They vowed to remain for two years, recycling water, air, and waste and growing their own food, to prove that a human colony could survive on another planet.

Within weeks, they were gasping for air, crops had failed, and the crystal clear "ocean" had turned to slime. It would later be revealed that air and food had to be smuggled in, and scrubbers were used to reduce the dangerous buildup of carbon dioxide. Even so, the already trim biospherians emerged two years later an average of twenty-five pounds lighter. This elaborate sealed-off environment, far grander than anything that could conceivably be transported to Mars, had been unable to sustain eight humans.

Zubrin is undaunted by such failures. His vision goes far beyond life under a dome. He plans for the day when humans will melt the polar caps of Mars, which consist largely of frozen CO_2. This will create a greenhouse effect. Mars will once again be warm and wet. Plants will thrive in the CO_2 atmosphere, freeing oxygen. An atmosphere of breathable air will develop, and the colonists can put aside their space suits. Zubrin speaks of creating an Earthlike atmosphere on Mars, even though at present hundreds of scientists, using the latest techniques, are struggling to understand the forces that determine Earth's atmosphere. But for Zubrin, these are details. All we need to make it happen is the will. Much was said in Boulder that summer about "destiny."

The trouble with the future, someone said, is that there are so many of them. The elaborate futuristic scenarios of "visionaries" like O'Neill and Zubrin have a powerful appeal for those who have fallen in love with technology. But as we seek to invent the future, we should be prepared to adjust our dreams to new realities revealed by science.

INVENTING THE FUTURE

Arthur C. Clarke, who is probably best known as the author of *2001: A Space Odyssey*, predicted in a 1945 article in *Wireless World* that artificial satellites in geosynchronous orbits would one day be used to relay radio messages around the world. A satellite in a geosynchronous orbit, which is at an altitude of about twenty-three thousand miles, has an orbital period of exactly twenty-four hours, just matching the rotation of the Earth. To an observer on Earth, the satellite thus appears to remain stationary. Communications experts scoffed; in 1945 the idea of an "artificial moon" was still science fiction. It would be another twelve years before the Soviets would shock the world with the launch of *Sputnik I*.

It was a brilliant insight. Today, there are nearly two hundred communications satellites; it's a $15 billion per year business and still growing, but it's doubtful that communications satellites as envisioned by Clarke would have been practical. His satellites were manned space stations, with living quarters for a crew whose principal task was to replace vacuum tubes as they burned out. Arthur C. Clarke foresaw communications satellites, but he did not foresee microelectronics—no one did. Just two years after he described his dream of space stations, the transistor was invented, and soon after, the integrated circuit. No larger than Volkswagens, each of today's communications satellites flawlessly relays millions of times as much information as the huge manned space stations Clarke proposed—and today's satellites have no need for a crew.

Science is a wild card. The further we try to project ourselves into the future, the more certain it becomes that some unforeseen, perhaps unforeseeable, advance in science or technology will shuffle the deck before we get there. Often, as in the case of semiconductor electronics, science provides us with a future far beyond our dreams; other times it reveals unexpected limits. Science has a way of getting us to the future without consulting the futurists and visionaries.

The historian Arnold Toynbee once explained his phenomenal productivity: "I learn each day what I need to know to do tomorrow's work." Science advances in much the same way. With each hard-won insight, the scientist pauses just long enough to plot a new course, designed to take advantage of what has just been learned. Before some distant goal can be realized, new discoveries may render it less desirable or reveal a more attractive alternative. Science keeps offering new futures to choose from and crossing old ones off the list.

Politicians, by contrast, want to agree on a future and establish policies that will get us there. In the exploration of space, this has led to an ever-widening disconnect between scientists and politicians as they pursue fundamentally different objectives. Sixteen nations have been committed by their politicians to the International Space Station, a scientific project that is scorned by their scientists.

THE VIRTUAL ASTRONAUTS

Meanwhile, far beyond the orbit of Pluto, the most distant planet in the solar system, *Pioneer 10* has long since completed its fantastic voyage of discovery among the outer planets. The first spacecraft to venture beyond Mars, *Pioneer 10* navigated the unknown hazard of the asteroid belt to send back the first close-up images of the giant planet Jupiter. The tiny 570-pound spacecraft went on to chart the currents of the solar wind all the way to the edge of interstellar space, while surviving on less energy than it takes to operate your porch light. Launched in 1972, the same year that *Apollo 17* flew the last mission to the Moon, *Pioneer 10* was more than six billion miles from Earth. The Sun from that distance would look to human eyes like just a bright star.

Built for a two-year mission, *Pioneer 10* suffered the usual infirmities of old age: its mechanical limbs became arthritic; its senses were dimmed by the battering of radiation and micrometeoroids. Some of its circuits had been shut down to conserve energy; the nuclear furnace that supplied its power was slowly growing cold. Nevertheless, the little spacecraft still answered the phone, faithfully reporting back measurements of the last traces of the solar wind. In April 1997, however, *Pioneer 10* was passed by a younger, faster Voyager spacecraft. It was no longer needed; scientists who had spent twenty-five years analyzing data sent back by *Pioneer 10* hung up the phone for good. It did not matter that *Pioneer 10* would never return; it was expendable. It is *Pioneer 10*, not the Apollo program, that is the paradigm for the future of space exploration.

The exploration of the solar system began with the *Mariner 2* flyby of Venus in 1969. Information sent back by *Mariner 2* and later American and Soviet missions ended any thoughts of a human expedition to Venus. There are no swamps on Venus; clouds, not of water but of sulfuric acid, hide a surface hot enough to melt lead. In 1990 the *Magellan* spacecraft would use radar eyes to map the entire surface of Venus—a surface no human eye could ever see.

Less than a month after the *Apollo 11* lunar landing, *Mariner 6* and 7 returned the first close-up pictures of Mars. There were no canals; the desiccated surface was pockmarked with craters like the surface of the Moon. It did not appear to be the surface of a living planet, but there were vast arroyos showing that great torrents of water had once coursed across the face of Mars. In 1976 two Viking landers set down on the Martian surface, sending back pictures of a harsh rock-strewn landscape. They even tested the Martian soil for traces of life. No clear evidence of life was found. It would be twenty-one years before *Pathfinder* would land on Mars carrying the Sojourner.

The tiny robot caught the imagination of people everywhere. It was not seen as some inanimate object but as an extension of ourselves; we were all on Mars. Its brain was the brain of humans back on Earth. Its senses were the senses that humans gave it. Sojourner even had a sense of smell, an atomic "sniffer" that allowed it to analyze the composition of the rocks it found. Thermocouples felt the midday warmth of the sand beneath its wheels and the frigid cold of the thin atmosphere just a couple of feet higher. Slow but steady, the little robot did not break for lunch or complain about the cold nights. Sojourner was the first of a new breed of rover telerobots that will give scientists a virtual presence in places no human could ever venture.

As with *Pioneer 10*, when its job was done it did not matter that there was no way to return. The total *Pathfinder* mission to Mars, 150 million miles from Earth, cost only about a fourth as much as a single launch of the space shuttle into low-Earth orbit. The *Galileo*, having completed its primary mission to Jupiter, is now studying the Jovian ocean moon Europa, and the *Cassini* is making the long journey to Saturn and its moon Titan. Robot telescopes are daily revealing new wonders at the very edge of the universe. These telerobots are simply extensions of our frail human bodies. And with every day that passes, we learn to build better robots. Humans, by contrast, haven't changed much in thirty-five thousand years.

These are all the things, if there had been time, that I would have liked to explain to the members of the Space Subcommittee at the hearing on the International Space Station. Perhaps if they understood where we've been, they could understand where we are. The scientific accomplishments of the astronauts on board the the space station will be inconsequential. It is the scientists who control the robots, having become virtual astronauts, who are exploring the universe.

THE TIME TRAVELER

In November of 1998, after an absence of thirty-six years, John Glenn returned to space as a member of the crew of the space shuttle *Discovery*. NASA's insistence that Glenn was being sent back into space for scientific reasons was not taken seriously by scientists, the media, or the American people. Yet few begrudged Glenn his nostalgic return to space. His first journey into space had restored the self-confidence of the nation. From his position in the U.S. Senate, he had lobbied hard for the chance to end his long career of public service with a second trip into space, and most Americans thought we owed it to him.

Just one week before John Glenn's return mission, NASA announced that the seventy-seven-year-old astronaut had been dropped from a "high-priority" age-related experiment for medical reasons. The experiment was to study the effect of the hormone melatonin on the adjustment of a seventy-seven-year-old to the ninety-minute day-night cycle on the shuttle. The effect on younger astronauts, it seems, had already been studied on previous missions.

The medical reasons for dropping Glenn from the experiment were never disclosed, but the more puzzling question was why it had been considered a high-priority experiment in the first place. Not much about the aging process can be deduced from the response of a single seventy-seven-year-old subject, and if the object was to learn about adaptation to ninety-minute day-night cycles, they hardly needed to launch the shuttle. They could have rented a motel room, drawn the blinds, and installed a twelve-dollar electronic timer on the light switch. The full cost of the shuttle mission was about $1 billion—roughly two years of funding for the hundreds of serious peer-reviewed extramural research grants awarded by the National Institute on Aging.

John Glenn's return to space on board *Discovery* was uneventful. There was a rather long delay in leaving *Discovery* after it landed at the Kennedy Space Center; Glenn had difficulty standing after eight days of weightlessness, but the other astronauts waited patiently to leave the shuttle as a group. There has been no announcement of any scientific results from the trip. Glenn was once again given a parade down Broadway, but this time the crowd along the parade route seemed not to be much more than the usual lunchtime throng of New Yorkers. A few stopped to watch and wave as the motorcade went by.

At a press conference, Glenn spoke of the changes that had taken place in the space program since his last trip; the space shuttle, of course, is a far cry from the cramped Mercury capsule. But the real symbolism of the mission was that after thirty-six years, John Glenn had traveled only eighty miles further from Earth than he did the first time. America's astronauts have been left stranded in low-Earth orbit, like passengers waiting beside an abandoned stretch of track for a train that will never come, bypassed by the advance of science.

FIVE
THERE OUGHT TO BE A LAW
In Which Congress Seeks to Repeal the Laws of Thermodynamics

COLD FUSION GETS ITS DAY
IN CONGRESS

BY 8:30 A.M. ON APRIL 26, 1989, a long line snaked down the marbled hall of the Rayburn House Office Building, waiting for the doors of the Science Committee hearing room to open. The hearing on "recent developments in fusion energy research" was not scheduled to begin for another hour. Network camera crews were already setting up inside, and racks of their electronic equipment were presided over by technicians in the hallway. Most of those in line would be turned away; the hearing room only has about sixty chairs, many of which would be reserved for witnesses in what was to be an all-day hearing.

The "fusion pioneers," as Stanley Pons and Martin Fleischmann were called by the *Salt Lake Tribune*, were to be the star witnesses. Barely a month had passed since their announcement of the discovery of cold fusion. They seemed like an odd couple, the urbane sixty-two-year-old Fleischmann, with his not-quite-identifiable European accent, and the younger, nerdish-looking Pons, whose accent betrayed his roots in a small rural town in the hills of North Carolina. Elected to the Royal Society, a very high distinction in the United Kingdom, showered with honors, Fleischmann radiated brilliance. Ideas on every subject seemed to spew forth effortlessly from his fertile brain, but few of his ideas were practical. Pons had been a student of Fleischmann's at the University of Southampton in England. What Pons lacked in brilliance, he made up in aggressiveness and energy.

But despite the superficial differences, they were too much alike to be effective collaborators. Neither had much taste for slow, careful science. They were both scientific gamblers, given to playing long shots. Together they could be almost manic. It is a pattern we will see repeated. The junior collaborator, dazzled by his famous mentor, believes him to be incapable of error. The older is confident that if he has made a mistake, his clever young friend will surely catch it. Thus are self-doubts suppressed as they carry each other along on an intoxicating ride.

They arrived at the hearing room with a full delegation from the University of Utah, including the president, Chase Peterson. There was the usual milling about the witness table as introductions were made. Often only a handful of committee members are present for hearings, but Science is a very large committee, and on this day it appeared that all forty-eight members were there for the start. The hearing was finally gaveled to order by the chairman, Robert Roe of New Jersey, at 9:45. Noting that press reports of attempts to duplicate the Utah results were conflicting, Roe said the objective of the hearing was "to allow an interchange among experts with different views and to allow members of the committee to assess the significance of current information."

In the opening remarks by committee members, Representative Bob Walker, a Pennsylvania Republican, set the tone. He proposed that the $5 million that had been redirected to cold fusion be increased to $25 million. It is, he said, the least we can do. Representative Wayne Owens from Utah declared that cold fusion was nothing less than a miracle. The rest of the morning was reserved for the University of Utah to make its case. The Utah show was stage-managed by Cassidy and Associates, a Washington lobbying firm that was famous for securing huge pork-barrel appropriations for client universities. They even brought along an architect with drawings of what a cold fusion power plant might look like—complete with parking lots and a cafeteria.

Congressional hearings are theater, intended less to gather information than to give the committee members a stage on which to make statements. The walls of the high-ceilinged Science hearing room bear life-size portraits of past committee chairmen. The committee is elevated on tiers, like those in an amphitheater, so that the witnesses look up at the congressmen. The committee chairman is enthroned in the center of the top tier. When called to testify, witnesses are seated behind a heavy table that seems to be an inch or so higher than a normal table, making them feel that they have been shrunk in stature. Invariably, as they begin speaking, the chairman stops them and tells them to speak directly into the microphone, which because of the high table must be pulled down. The whole atmosphere is designed to impress witnesses with their insignificance before this powerful body.

Fleischmann and Pons were the first witnesses. Pons appeared somewhat subdued by the proceedings, anxious to get through his prepared remarks and turn the microphone over to Fleischmann, who, by contrast, was in his element. Neither admitted to the slightest doubt concerning their discovery. The softball questions that followed from the committee were generally prefaced with congratulations. The only probing question dealt with the discrepancy between the neutron levels and the excess heat. Fleischmann replied that he would rather not comment on that. The implication seemed to be that an answer might involve proprietary information, and no one pressed him any further on the subject.

Next up was the president of the University of Utah. Chase Peterson had made the remarkable leap from director of public information to the presidency by virtue of his deft handling of the publicity surrounding the Jarvik artificial heart, which was developed at Utah. You may recall the tragic figure of Barney Clark, the sole recipient of the Jarvik heart, whose last days on Earth were a man-made Hell somewhere between life and death. As the nation followed every development in the case, it was Chase Peterson, an M.D. by training, who would appear each day before the television cameras in a doctor's white smock with a stethoscope hung about his neck to update the world on Barney Clark's condition. It appeared as if he had just come from the patient's bedside, although he had not practiced medicine in years. He played his role to perfection. The Jarvik heart was a medical failure, but it succeeded in establishing Utah as a major player in advanced medical procedures. Peterson had reportedly recalled that outcome in urging the release of the cold fusion announcement.

In making his pitch for the $25 million, Peterson emphasized that the State of Utah had already committed $5 million to the project. He also claimed that the university had raised over $1 million in "private funds." He had, in fact, earlier used the claim of a large "private donation" to persuade the Utah legislature to commit the $5 million. A year later it would come out that the "private donation" was actually from a secret university account controlled by Peterson, not from an outside source as he had claimed. When the source of the money eventually became known, Peterson was censured by the faculty and forced to resign in disgrace.

But that was far down the road. On this spring day in Washington, Peterson was accompanied at the hearing by a "business strategy consultant" named Ira Magaziner, who would show up a few years later as an advisor to President Clinton. It was Ira Magaziner who caught the imagination of the Science Committee. Magaziner urged Congress not to "dawdle around" waiting for the scientific community to confirm cold fusion, or the Japanese would beat us to commercialization. His argument was simple. He admitted that he knew nothing about the science, but if cold fusion held up, it could spawn the biggest industry in the history of the world, worth hundreds of billions of dollars. If it failed, the country would be out a paltry $25 million. This was language the committee understood. The only witness with no science background had the greatest impact on the committee.

It is the same argument that leads people to stand in line for hours to buy lottery tickets when the pot reaches record levels. The danger, of course, is that the wish can be so powerful that you are tempted to wager on the impossible. In his book on cold fusion,

Gary Taubes calls this "Pascal's wager." Blaise Pascal was a renowned seventeenth-century physicist and mathematician who at the age of thirty-two renounced a life of science for one of faith. "Do not hesitate to wager that God exists," he advised. "If you win, you win everything." Needless to say, they love Pascal in Las Vegas. Some form of Pascal's wager is often invoked to justify impossible schemes. As we wander through the world of voodoo science, we will be on the lookout for Pascal's wager in its many guises.

There were many other witnesses to be heard, but as the University of Utah completed its testimony, there was a stir in the hearing room: the television cameras were being trundled out. Chairman Roe turned the gavel over to Representative Marilynn Lloyd of Tennessee and left, along with most of his colleagues; many spectators and reporters followed the television cameras. They came to see the stars, and the stars were off the stage. The dogwoods were in bloom in Washington, and the day belonged to cold fusion.

On May 1, just five days after the cold fusion hearing before the House Science Committee, a special evening session on cold fusion was held at the Annual Spring Meeting of the American Physical Society in Baltimore. Pons had agreed to be one of the speakers, but he canceled his appearance at the last minute, explaining that he was too busy preparing for the visit of a planeload of congressional dignitaries, arranged by Jake Garn, the conservative Republican senator from Utah.

Meanwhile, I learned from a colleague at the University of Utah that Pons was not in Salt Lake City preparing for a congressional visit at all. The rumor was that Pons was in Washington, D.C., only forty miles from Baltimore, for a meeting with White House Chief of Staff John Sununu, and perhaps the president himself. Confirming that rumor produced one of those wonderful moments that make Washington such an interesting place to work. I took the direct approach of simply calling Sununu's office and asking, "Is Governor Sununu going to meet with Professor Pons?" "I cannot confirm that," said the very official voice at the other end, "since the meeting is confidential." Pons was expecting to meet with Sununu on May 3.

On May 2, however, the papers carried the story of the cold fusion session in Baltimore the evening before. It had not gone well for cold fusion: theorists had reported that cold fusion violated not one but several accepted physical principles; chemists seemed to be able to account for all of the heat without invoking nuclear reactions; elementary flaws in the Utah experiment were laid bare. Moshe Gai presented the results of the Yale-Brookhaven collaboration, showing that neutron levels were not merely too low— there appeared to be no emission at all. A scientific consensus was taking form. Newspapers gave the public its first real indication that something seemed to be seriously wrong. Sununu called Allan Bromley, the president's nominee for science advisor, and asked for an assessment of the Baltimore meeting. Citing "urgent matters" that had arisen, Sununu canceled his meeting with Pons.

The day after the Baltimore meeting of the Physical Society, I was again interviewed by NBC's Bob Bazell, this time on the *Today* program. Cold fusion, I told the audience, is dead, but the corpse won't stop twitching. Inept scientists who had rushed to report confirmation, greedy university administrators who had tarnished the reputations of their institutions, gullible politicians who had wasted the taxpayers' dollars, and careless journalists who had accepted every press release at face value: all had an interest in pretending the issue had not been settled. I never imagined, however, that a decade later there would still be scientists championing cold fusion, or that companies claiming to have developed cold fusion devices would attract investors.

The turnaround on Capitol Hill was dramatic. Physicists who had been involved in the Baltimore session on cold fusion met the next day with stunned members of the House Science Committee, who just a few days earlier had proudly posed with Pons and Fleischmann for press photographers. Representative Walker, annoyed by what he took to be arrogance on the part of the skeptical physicists, refused to withdraw his amendment to move $5 million from high-temperature fusion to cold fusion, but it was never brought to a vote. The "Garn Express" to Salt Lake City was first postponed, then quietly canceled, as congressmen who had signed up for the trip discovered reasons why they couldn't make it. "Cold fusion?" scoffed Washington comedian Mark Russell. "In Salt Lake

City you can't even get a cold beer." Cold fusion was becoming a joke. In Washington, that's usually fatal.

The hearing had failed to alert the Science Committee to the extent of scientific skepticism about cold fusion. But this was not the first time a congressional committee had been lured by the promise of free energy. Just three years earlier, Joe Newman, the backwoods genius who claimed to have invented the Energy Machine, had his day in Congress.

JOE NEWMAN'S DAY IN CONGRESS

Before the advent of air conditioning, British diplomats stationed in Washington received the same hardship pay as those assigned to Calcutta. Congress still follows a tradition set in that earlier time and closes down for the month of August. The summer of 1986 seemed particularly hot and humid. On July 29, I got a call from the Senate Government Affairs Committee. Senator Thad Cochran of Mississippi had scheduled a hearing for the next morning on legislation to force the Patent and Trademark Office to issue a patent for "some sort of generator." He hoped to bring the legislation to the floor before Congress left town. It promised to be a controversial hearing, and my caller thought somebody from the scientific community should be there as an observer. Would I be willing to sit in on it? "By the way," he added, "the inventor will be there to testify. His name is Joseph Newman." I had almost forgotten about Joe Newman. The humble mechanic from Lucedale had come a long way; he would have an opportunity to make his case before a powerful committee of the United States Senate.

When we left Joe Newman in chapter 1, he had the crowds in the New Orleans Superdome swaying in their seats. His suit against the U.S. Patent and Trademark Office seemed to be going his way as well. Against the testimony of Patent Office experts, Newman had presented his own experts—none other than physicist Roger Hastings and engineer Milton Everett, the same pair that had vouched for the Energy Machine on CBS News.

Newman argued that the Energy Machine was not a perpetual motion device, and therefore he should not be subject to the 1911 policy of refusing patents for such devices. He claimed that the energy to run his machine came from the conversion of mass into energy according to Einstein's famous $E = mc^2$. Slowly, Newman said, his machine was devouring its own copper wires and iron magnets. Because c^2 (the speed of light squared) is such a huge number, his machine would, for all practical purposes, last forever.

It was an extraordinary claim. Curiously, $E = mc^2$ had resolved the first serious challenge to the conservation of energy—Becquerel's discovery of radioactivity in 1896. French physicist Henri Becquerel discovered that certain minerals continuously radiated energy. Some were actually warm to the touch and glowed in the dark. They seemed to keep emitting energy indefinitely. Where was the energy coming from? The very foundations of science appeared threatened. The answer came in 1905 when Albert Einstein produced the *special theory of relativity*. It's worth taking just a minute to discuss Einstein's theory and the relationship between mass and energy.

Einstein did not set out to solve the puzzle of radioactivity. He was fascinated by light. It had been known since the experiments of the French physicist Augustin-Jean Fresnel in 1815 that light behaved like waves. But waves in what? Unable to imagine waves in a vacuum, scientists had invented the "ether," a substance that was imagined to fill the universe. Light, it was supposed, must move through the ether in the same way that ripples move over the surface of a pond. Therefore, it should be possible to determine the motion of the Earth through the ether by comparing the apparent speed of light in the direction the Earth is moving with the speed of light at right angles to the direction of motion. But when A. A. Michelson and E. W. Morley actually tried the experiment in 1887, the speed of light was found to be the same in all directions. It was as if the Earth was sitting motionless in the ether with the sun and stars rotating about it, just as the Catholic Church had insisted to Galileo in 1633. That was not a comfortable thought for physicists.

Einstein proceeded to derive the mathematical transformation of time that would result in the measured speed of light being the same no matter where you are or how you're moving when you make the measurement. That transformation is known as the theory of special relativity. When Einstein used this transformation to calculate energy, he got a stunning result: an object has an energy $E = mc^2$ even when it's stationary. The theory of special relativity had revealed the equivalence of mass and energy. The implications of Becquerel's discovery of radioactivity were now clear: the law of conservation of energy was not overturned, it was unified with the conservation of mass, which had been thought to be an entirely separate law of nature.

We would never think about the universe in the same way again. A thousand mysteries were swept away at once. Energy is forever being converted to mass and then back again. When you wind your alarm clock at night (of course, no one really winds up alarm clocks anymore) you actually increase its mass, ever so slightly, by the energy stored in the deformed chemical bonds of the spring. As the clock ticks, the spring unwinds, converting this imperceptible amount of mass back into energy. The energy takes the form of heat, generated by the friction of the clock works, and sound waves, generated by the ticking. A physicist would say that when the spring is fully unwound, the clock is in its "ground state"—the state of lowest energy. The clock weighs less in its ground state, but the difference is much too small to measure.

The nucleus of an atom can be in a state of excess energy, like the wound-up clock. Such a nucleus is said to be radioactive. There is a statistical probability that a radioactive nucleus will abruptly "unwind," giving up its excess energy all at once and becoming a different, and somewhat lighter, nucleus. The energy is carried away by the sort of atomic radiation Becquerel had discovered. For example, a helium atom formed by fusion, discussed in the first chapter, weighs slightly less than the sum of its parts. The difference, or "mass defect," is converted into energy according to the Einstein formula, $E = mc^2$. The energy is carried off by nuclear radiation. But the mass defect is only a small fraction of the total mass. Was it possible to convert entire atoms into electrical energy as Newman claimed? And without producing any radiation? Could Joe Newman, a simple backyard tinkerer, have discovered a way to do it?

Judge Thomas Penfield Jackson of the U.S. District Court of the District of Columbia, acknowledging his own technical limitations, referred the matter to a special master, electrical engineer and former commissioner of patents William E. Schuyler Jr. To the utter dismay of the Patent and Trademark Office, Schuyler's report to the court on September 28, 1984, concluded that the evidence was overwhelming that the output of the Newman Energy Machine was greater than the input. Newman's supporters were jubilant.

Judge Jackson, however, was unpersuaded. Perhaps he had simply learned that in this life you never get something for nothing. Relying on his own common sense, the judge set out to learn a little physics. Eight months later, on June 11, 1985, Judge Jackson held that Schuyler's report was clearly erroneous. In support of his position, the judge cited the laws of thermodynamics and a report from Mississippi State University. He set aside the report of William Schuyler, the special master he had himself appointed. Instead, he ordered Newman to turn his Energy Machine over to the National Bureau of Standards (NBS), perhaps the most trusted laboratory in the nation, for testing. Newman and his lawyers, who felt they had won, bitterly complained that the judge's action was unfair.

Several members of Congress apparently agreed. In the spring of 1986, Representative Bob Livingston (R-LA), the chairman of the Republican Study Committee, circulated a special report, "The Patent Office and Joseph Newman: An Abuse of Power." The report concluded that Joseph Newman had "received arbitrary and unfair treatment at the hands of the Patent and Trademark Office." Six members of Congress had already been persuaded to submit "private relief" bills to force the Patent Office to issue Newman a patent for "an unlimited source of energy." The six included both senators from Mississippi, Thad Cochran and Trent Lott. Lott would one day become Senate majority leader and aspire to the presidency of the United States.

I stayed up late that night reading and then rereading *The Energy Machine of Joseph W. Newman,* a book written and published by Newman. The book was a passionate tangle of simplistic philosophy, boastful autobiography, and garbled scientific ideas, all leading up to his Energy Machine.

Newman ran away from an orphanage at fourteen to make his own way in the world. Self-taught, he had a genuine aptitude for mechanics. Like countless schoolboys, Newman was fascinated by electromagnets. A simple loop of wire carrying an electrical current produces a magnetic field at its center proportional to the electrical current. If you make two loops of the wire, the magnetic field is doubled, even though the current remains the same. This hit Newman like a divine revelation. You could create as large a magnetic field as you wished by just adding more turns—with the current supplied by a single battery! That's true, of course, but there is a price to pay. When you connect the battery, the growing magnetic field tries to induce an opposing electrical current, called a back-electromotive force, or just "back-EMF." The result is that the more turns you have, the more the current is impeded, and the longer it takes for the magnetic field to build up to its full value. This is known as Lenz's law; it is the conservation of energy applied to electrodynamics.

Apparently Newman did not get that far in his self-help course. He convinced himself that the more turns you put on an armature, the more efficient an electric motor should be—until finally you would get more energy out than you put in. He imagined that he had made a profound discovery that had eluded everyone else. What he ended up with was a motor that ran on a very small current—but a huge voltage was needed to supply that current. Since the power drawn by the motor is given by the product of the current and the voltage, the increase in voltage, as you would expect from the law of conservation of energy, just offsets the lower current. There is no increase in efficiency.

When Joe Newman told the Superdome crowd that his Sterling sports car ran on the current of a single battery, back in chapter 1, he was at the very least being evasive. The machine delivered to NBS was not powered by a single nine-volt transistor-radio battery, as Newman seemed to imply, but by 116 of the batteries connected in series, in a special battery pack provided by Rayovac. The same current flowed through all 116 batteries, but because they were connected in series, the total voltage supplied to the Energy Machine was about a thousand volts. Because they are inconvenient to operate—requiring increased insulation, for instance—high-voltage motors have few practical applications. And, of course, they cannot supply more energy than they consume. Newman, it appeared, had simply rushed ahead with his idea without fully understanding the physics. The whole business about $E = mc^2$ and the Energy Machine slowly devouring itself seemed to have been grafted on to his discussion at some later time to answer the criticism that his machine violated the conservation of energy.

I arrived an hour early the next morning for a 10:00 A.M. hearing, but the hearing room was already filling up with men in three-piece suits carrying briefcases. I took a seat near the back and listened in on the conversations going on around me. These, I soon discovered, were executives from major American corporations. Their reasoning seemed to be that if the Newman Energy Machine warranted a Senate hearing there might be something to it—and if the gravy train was about to pull out, these guys weren't going to be left standing on the platform. It was the same reasoning that Ira Magaziner had used to urge funding for cold fusion. It was Pascal's wager.

Then Joe Newman strode in, surrounded by a small entourage. Among those who had come into the room with Newman, I recognized Roger Hastings, the physicist who had vouched for Newman on CBS and testified for him in court. The buzz of conversation abruptly quieted; every eye was on Newman. The mechanic from Lucedale had been transformed since I saw him on the CBS *Evening News* thirty months earlier. Carefully groomed and dressed in a three-piece suit like the captains of industry that filled the room, he was treated with obvious deference by the people that crowded around him. One thing had not changed—whatever charisma is, Joseph Newman still had it.

The hearing began with those members of Congress who had introduced legislation on Newman's behalf going over the history of the unfair treatment Newman had been subjected to in his efforts to obtain a patent. Applause from Newman supporters in the audience led Senator Thad Cochran, who was chairing the hearing, to caution the audience against any reaction to the testimony of witnesses.

Seated in the front row, which was reserved for witnesses, I recognized John Lyons, the director of the National Engineering Laboratory of the National Bureau of Standards. As soon as the congressmen had their say, Lyons was called to the witness table.

In brief, understated testimony Lyons explained the procedure used by the NBS scientists to test the Energy Machine. He also revealed a whole new aspect of the controversy; Joe Newman, its seems, was already well known to the NBS laboratory.

In 1975 Newman had written to the Office of Energy Related Inventions at NBS asking for an evaluation of his invention of an "unlimited source of energy." The office had responded by asking for more details, including test results, since Newman's claims ran "contrary to well-established scientific principles." Newman never provided the additional information.

Seven years later, however, in 1982, he showed up unannounced at the campuslike NBS laboratory in Gaithersburg, Maryland, with his Energy Machine in tow, pleading with NBS to test it. Exasperated NBS officials turned him away again, since he was still unable to explain in any coherent fashion the principles on which his machine was based.

Why then, in 1986, would Joe Newman fiercely resist a court order to deliver his Energy Machine to NBS and seek to obstruct the very test he had begged NBS to conduct four years earlier? If we could answer that question, it might help us to understand the key question in voodoo science. Few scientists or inventors set out to commit fraud. In the beginning, most believe they have made a great discovery. But what happens when they finally realize that things are not behaving as they believed?

Physicists had suggested to Newman that he try connecting the output of the Energy Machine to the input; if the machine worked, he should be able to do away with the batteries. Perhaps sometime between 1982 and 1986 he followed their suggestion and discovered that his machine did not work. However it happened, the line between foolishness and fraud must have been crossed.

Lyons concluded his testimony with the classic understatement that "the input power from the batteries running Mr. Newman's device exceeded the output power from the device." Newman followed Lyons to the witness table, accompanied by his lawyer and Roger Hastings. He seemed unperturbed by Lyons's testimony. This was, after all, just the sort of debate with a Ph.D. scientist he had always said he wanted, although Lyons was a Harvard chemist and not a physicist. Moreover, the debate would be judged by a panel of U.S. senators, a group more comfortable with Joe Newman's revival-tent style than with the First Law of Thermodynamics. Newman began his testimony with an apparent reference to Judge Jackson's decision to set aside the report of William Schuyler, the expert the judge had himself appointed to examine Newman's claim. "I will try to keep my anger down," he told the committee, "because I am boiling over. When I see in front of a courthouse a woman blindfolded holding up the scales of justice, that touches me, and when I see injustice, it makes me boil like a damn tornado." His powerful voice trembled with indignation, "I'm not just fightin' for Joe Newman." Newman finished his emotional testimony with his standard challenge to any Ph.D. physicist to debate him.

It was Senator John Glenn of Ohio, former astronaut and genuine American hero, who rather reluctantly took over the questioning of Newman. His calm voice was that of the test pilot and astronaut, always in control of his emotions. Glenn admitted he was no Ph.D. physicist, but it would turn out that he had all the technical background he needed for this debate. "It's a simple enough problem," Glenn said. "You measure the input and you measure the output and you see which is larger. Would Mr. Newman agree to that? If he does," Glenn went on without pausing for an answer, "what laboratory would he like to have make the measurements?" Newman's bluff had been called. It was the first time I had seen him speechless. Pressed to respond, he finally said he objected to any tests by any laboratory on the grounds that it would be an affront to the scientists who had already vouched for his machine. You could feel the room stiffen. What kind of answer was that?

A member of the committee staff handed Senator Glenn a note. Glenn looked up from the note and asked one more question. Had Newman known William Schuyler before the trial? The room went silent. What was behind Glenn's question? Schuyler's report was the most troubling aspect of this entire affair. If there was any link between Newman and Schuyler, it would deprive Newman of his claim that by setting aside the special master's recommendation Judge Jackson had treated him unfairly. Clearly flustered, Newman acknowledged that he had met Schuyler once, but he said Schuyler didn't even remember him. In fact, Glenn pressed, hadn't Schuyler's patent-law firm once represented Mr. Newman? "Yes, but Mr. Schuyler did not know me personally." The only sound in the hearing room for a moment was a sort of collective sigh as this revelation sank in. Feeling the mood turning against him, Newman resorted to bluster: "I see where you're headed, Senator. I don't have anything to hide. Look me in the eye, you won't see me blink." "I ain't blinkin' either," Glenn replied evenly.

An alert member of the Senate staff, going through a mountain of documents related to the Newman case, had spotted Schuyler's name on the letterhead of a patent firm that represented Newman in an earlier invention. The hearing went on for another two hours with testimony from Newman's supporters and from the Patent Office, but the audience began drifting away. They understood that the exchange between Joe Newman and Senator Glenn had put an end to congressional intervention on behalf of Joseph Newman and the Energy Machine.

The great irony is that it was not the towering authority of the First Law of Thermodynamics that brought down the Energy Machine. Aside from Senator Glenn, it's not clear that anyone on the Committee ever quite understood what the conservation-of-energy argument was all about. Most members of Congress, after all, are lawyers. What lawyers can recognize a mile away is a conflict of interest. Congress had been spared the embarrassment of legislating a patent for "an unlimited source of energy." It would not, however, have been the first such embarrassment for Congress; there are remarkable parallels between the case of Joe Newman and that of Garabed Giragossian sixty seven years earlier.

GARABED GIRAGOSSIAN'S DAY IN CONGRESS

In the autumn of 1917, an Armenian immigrant, Garabed Giragossian, announced that he had discovered a machine that produced more energy than it took to run it. Like Joe Newman, Giragossian had no scientific training but was endowed with boundless self-confidence and energy. Moreover, he had his own circle of technologically unsophisticated admirers who fed the press glowing testimonials. The press did for Garabed Giragossian what CBS television would do for Joe Newman: they played up the story of a self-educated immigrant genius whose invention confounded the pompous experts who had declared it couldn't be done—and all Garabed wanted was to be certain it would be developed by his adopted homeland. It was the American story, and the public loved it.

Giragossian also had his day before the U.S. Senate. He declined to reveal the details of just how his machine worked, but he proposed that the president of the United States personally appoint a team of top engineers and scientists to examine it. A number of physicists wrote to Congress, warning that the Giragossian claim violated the laws of thermodynamics. But Congress overwhelmingly passed a special act calling for a presidential commission to look into Giragossian's discovery. After all, if Giragossian was right, imagine what it would mean for the nation and the world? It was again Pascal's wager and the lure of free energy.

President Woodrow Wilson himself, as Giragossian had proposed, appointed the distinguished team of scientists and engineers. Such high-level involvement fueled unrestrained press speculation about what the project might mean in terms of free electricity and factories without smokestacks. The public awaited the commission's report with high expectations.

It didn't take long. A demonstration was arranged. What the commission was shown was a flywheel, differing only in size from the flywheels that propel children's toy cars across a room. To get Giragossian's huge flywheel started, a muscular assistant operated a mechanical crank. Once it was started, however, the flywheel, which was mounted on bearings to reduce friction, was driven by a small electric motor. The distinguished commission watched as the flywheel slowly came up to its maximum speed. Then, using a dynamometer, Giragossian measured the energy required to bring the flywheel to a stop. The result, a beaming Giragossian proudly announced, was two hundred times as much energy as the electric motor supplied.

There was a shocked silence. They had been assembled for *this*? Giragossian had confused power and energy. He simply had not understood that energy, supplied by the muscles of his assistant and the electric motor, was being stored in the flywheel as it was brought up to speed over a period of minutes. When he abruptly brought the flywheel to a stop, all of the stored energy was expended in an instant.

The entire nation seemed embarrassed by the episode. There was no question of fraud; Giragossian was clearly sincere. It is ingrained in the American character to believe that a simple, virtuous man can accomplish things that are beyond the reach of closed-minded, so-called experts. People wanted Giragossian to be right, and the wish was made to seem plausible by credulous press reports. It is a lesson in why science has learned to insist on procedures that ensure openness. (We will learn more about the cost of secrecy in chapter 9.) Had Giragossian submitted his invention to the Patent Office or to scientific review, he could have been spared his eventual humiliation.

What is unusual about the Giragossian fiasco is how abruptly it ended compared to other examples of voodoo science. The reason, it would seem, is that it simply had not lasted long enough for an industry to build up around it. A few politicians and some members of the press may have been embarrassed, but there was no one, save Giragossian himself, who depended on the Giragossian project for a livelihood. It never had a chance to put down roots. There is something to be said for involvement of high-level scientific experts early on. The special presidential commission made quick work of what had become known as the Garabed project.

Prior to creating the presidential commission at Giragossian's request, there is no indication that Congress sought any advice from the scientific community or paid any attention to the unsolicited advice scientists offered. In defense of Congress, it might be argued that in 1917 Congress had little experience in dealing with technical issues. There are few issues before Congress today, however, that do not have a scientific or technological component. Where, then, does Congress turn for scientific advice?

THE DEATH OF THE OFFICE OF TECHNOLOGY ASSESSMENT

In the spring of 1995, Robert Walker (R-Pa.), elevated by the Republican takeover of Congress to chair of the House Science Committee, introduced The Hydrogen Futures Act. Its stated purpose was to promote the development of hydrogen, obtained from the decomposition of water, as a "new energy source." In principle, hydrogen is the perfect nonpolluting fuel; when it burns, the only combustion product is pure water. Walker's bill listed electric power generation as one of the potential uses of hydrogen and pointed out that, with most of the planet covered by ocean, the supply is inexhaustible. It is doubtful if more than two members of Congress understood that for hydrogen from the ocean to become a source of energy, it would first be necessary to repeal the laws of thermodynamics.

There is an unlimited supply of hydrogen in the ocean, all right, but it's in the form of H_2O. To obtain hydrogen as fuel, we have to separate the water molecules into hydrogen and oxygen. So why not extract the hydrogen from water, burn it as fuel, to generate electricity, and use some of the electricity to produce more hydrogen? You wouldn't even need to be near the ocean; when hydrogen burns, it recombines with oxygen to form more water, so you would only need an initial reservoir. Does this sound familiar? It should. It's a perpetual motion machine of the first kind—Robert Fludd's waterwheel in a slightly different guise.

It fails because the energy required to dissociate the water is greater than the energy that can be recovered when the hydrogen is burned. To prove this, the energies involved can be measured, and indeed they have been measured very precisely. But even without actually measuring the energies we know there will be an energy deficit; otherwise it would violate the conservation of energy. How many thousands of otherwise intelligent people have been separated from their money over the years because they never learned, never understood, or never believed the conservation of energy?

Robert Walker was a powerful member of Congress, chair of the House Science Committee. Was there no resource at his command that could have spared him this embarrassment? Ironically, at the time he submitted his bill, there was. Fourteen years earlier Congress had created the Office of Technology Assessment (OTA) to provide objective advice on scientific and technical issues. But this was 1995, the year of the "Contract with America." Before the year was out, OTA would be abolished in a symbolic demonstration that in downsizing the federal government Congress would not exempt its own bureaucracy. Robert Walker would cast one of the votes to eliminate OTA.

Many members of Congress were delighted to have the budget cuts to use as cover in eliminating OTA. It was, many complained, too slow for the pace of events in a rapidly changing world. It relied on the advice of panels of outside experts and often took many months to complete a study. This was a valid complaint. Cold fusion, for example, was all over in five months; OTA would have still been assembling its panel of experts. But the real problem went much deeper: Scientific reality frequently clashed with political goals. Indeed, there was a tendency to simply avoid seeking advice from OTA on controversial or partisan issues—which are generally the issues on which objective advice is most needed.

Congress, of course, is not unique in its fascination with infinite-energy schemes. There are, as I write this, at least three companies doing business in the United States that claim to have developed infinite-energy devices. These companies have attracted investments in the millions of dollars. How do they get away with it? For one thing, they often get a lot of free help from the media. We'll take a look at some of these companies, and the help they get, in the next chapter.

SIX
PERPETUUM MOBILE
In Which People Dream of Infinite Free Energy

THE RETURN OF JOE NEWMAN

"WELL, THE PACE OF technological change is breathtaking these days, isn't it?" Dan Rather asked rhetorically on the CBS *Evening News.* "Now a backyard tinkerer in Mississippi says he's built a machine, a kind of perpetual motion machine, that defies the laws of physics—and you know, some people think that just maybe he has. Bill Whitaker has taken a look at it." I couldn't believe it. It was March 11, 1987, more than three years since CBS first visited Joe Newman and his Energy Machine. Just when it looked like he had finally run out of energy, CBS was giving him another jump start!

Bill Whitaker caught up with Newman in Biloxi, Mississippi, standing beside the red Sterling sports car. The script had hardly changed from three years earlier. "It was hard for newsmen to know whether they were witnessing a technological farce or a genuine historic event," Whitaker began. "Joe Newman is the inventor of a controversial machine and has been saying for several years that it produces more energy than it consumes. Today, he hoped to show what it can do. He used it to power a car."

"Newman's Energy Machine didn't come from some high-tech laboratory," Whitaker tells the audience. "It came from his back-woods Mississippi workshop. Newman, a high-school dropout, built a contraption out of copper coils and spinning magnets that generates tremendous electromagnetic force from the current of a standard flashlight battery. He says it's based on the theories of Albert Einstein and will one day provide the world with an inexhaustible source of clean energy." Once again, the audience is fed the uniquely American myth of the self-educated genius fighting against a pompous, closed-minded establishment.

It was as if nothing at all had happened in the three years since CBS first visited Joe Newman. There was no mention of the failed test of the machine by the National Bureau of Standards; no mention of Newman's debacle before a Senate committee; no mention of the federal district court ruling against Newman in his patent suit. No mention either of the fact that the "standard flashlight battery" was keeping company under the hood of the Sterling with another 1,809 batteries.

"Most scientists are doubtful," Whitaker acknowledged. "They say Newman's invention defies the laws of physics. But a few who have examined the machine have become true believers." It was time for the "talking head"—the expert who would put it all in perspective. The head turned out to belong to one of the same "experts" that Bruce Hall had interviewed on CBS three years earlier. Milton Everett, the state highway engineer who had testified on Newman's behalf in the Senate hearing and in Newman's suit against the Patent Office, pulled out all the stops this time: "I think it's probably the most significant discovery in the history of man." Once again, as in the earlier CBS story, there was not even a token skeptic.

"Still," Whitaker said indignantly, "the U.S. Patent Office calls Newman's machine an impossible perpetual motion machine and, for eight years, has refused to grant him a patent." The camera turns to Joe Newman; it's now time for the inventor to speak for himself. Playing the role of the homespun genius perfectly, Newman looks squarely into the camera: "The best way to expose me is to issue the patent and throw me out in the public. If I'm wrong, nobody'd be embarrassed but Joe Newman."

The image on the screen flashes back to the red Sterling gliding to a stop. Newman emerges and waves to the crowd. "Today, Joe Newman was not embarrassed," Whitaker narrates in a hushed tone. "He brought his slow-moving car to a stop after two hours and said it could have gone on and on. Newman compared it to the first flight of the Wright brothers." The message could not have been clearer: those scientists who had bothered to examine Newman's machine were convinced. Moreover, Newman had tested his machine before the public with the CBS camera recording the event—and the Energy Machine had passed the test.

Why had CBS chosen to give a charismatic huckster access to millions of American homes, not once but twice? I called Bill Whitaker the next day to ask. "It was an update," he said defensively. "But you didn't update anything," I pointed out. "There was not a word about the Bureau of Standards test, or the court decision, or the Senate hearing." "That's not what the story was about," Whitaker protested. "It was about Joe Newman; this was a human interest story." That, it seemed, explained everything. After all, it was just entertainment. No one, unfortunately, bothered to explain that to the viewing audience.

Stories about fringe science are rarely held to the same standards as stories about politics, or foreign affairs, or even sports. Whitaker had begun by acknowledging that as a newsman it was hard to tell what was really going on. He was not at all embarrassed by his lack of scientific knowledge, nor did CBS seem to think it was necessary to send a reporter who knew about the conservation of energy to cover a demonstration of a perpetual motion machine. Try to imagine a reporter being sent to cover a football game who doesn't know that the tackle is an ineligible receiver. And while there's plenty of opportunity for opinion in reporting sports, things like the score are facts, and the media feels obligated to get that part of the story right. Those who watch football on television, after all, are generally very knowledgeable, and any slipup by a sportscaster is going to generate a flood of complaints. Relatively few people, however, feel confident enough about their own knowledge of science to challenge a science story. The media count on that.

Scientists tend to imagine that misreporting of science could be cured by having more reporters with a sound science background. It would surely help, but it's no guarantee. Consider the return visit of ABC News to James Patterson and his magic beads.

THE PATTERSON CELL REVISITED

There is, as we saw in the first chapter, a small but passionate subculture of true believers in cold fusion still residing on the fringes of the scientific community. Ignored or even ridiculed by other scientists, they dream of redemption when the world finally wakes up to the reality of cold fusion. There is even a magazine, with the unlikely title *Infinite Energy,* that fills its pages with rosy stories about the progress that's being made in cold fusion. On June 10, 1997, the editor of *Infinite Energy* sent me an e-mail to let me know that ABC's *Good Morning America* would carry an important story on cold fusion the following day. His message predicted that, because of its importance, the story would be in the first hour of the two-hour program.

I seated myself in front of the television at 7:00 A.M. hoping the cold fusion story would come up early in the hour, but the only thing resembling a "science" story in the first hour dealt with a report that the Loch Ness monster had been sighted in a lake in Turkey. After nearly two hours of the sort of gooey cheerfulness that is standard fare on early morning talk shows, *Good Morning America* host Charles Gibson was finally joined by ABC science correspondent Michael Guillen, who has a Ph.D. in physics from Cornell.

You may remember Guillen, the scientist turned TV reporter who interviewed inventor James Patterson, from chapter 1. Patterson claimed to be able to extract huge amounts of energy from ordinary water by passing current through a cell filled with tiny metal-coated beads. "We've gotten a tremendous response to that story," Gibson explains, "and Michael Guillen is back with an update." "Charlie," Guillen gushes, "it has come a long way in the last year, as you're about to see."

The tape rolls, and we see inventor James Patterson in his fishing boat reeling in a large-mouth bass. Who would not trust an avuncular, white-haired seventy-five-year old who loves to fish? The voice-over is Michael Guillen recounting Patterson's amazing discovery of a way to extract energy from water with his magic beads. Guillen still avoids using the term *cold fusion*. This, I learned later, was at the request of Patterson, who knows the term would cost credibility. But there is no denying the popular appeal of James Patterson, who complains that his wonderful discoveries are cutting into his fishing time.

The scene shifts from fishing to the gleaming new laboratories of Clean Energy Technologies, Inc. (CETI), the company created to commercialize Patterson's invention. It's a big step up from the cluttered garage workshop that was the setting for the program a year earlier. Clearly, Patterson and CETI have prospered—due in no small measure, I would imagine, to the free plug on *Good Morning America*.

"During the past year," Guillen informs the audience, "Patterson's beads have led to a big surprise. It turns out they also neutralize radioactivity." That would be a surprise, all right, to every nuclear physicist in the world. It would, in fact, be a miracle. The only way to "neutralize" radioactivity is to transmute radioactive isotopes into stable elements, something which, to the extent that it's possible, requires intense neutron bombardment from a nuclear reactor or a powerful nuclear-particle accelerator.

Patterson explains that when a solution containing uranium ions is circulated through the power cell, a device about the size of a roll of quarters, energy is generated and the radioactivity is gradually neutralized. "I'm skeptical," Guillen smirks. "Let's put it to the test." Patterson obliges, touching a button that starts the liquid circulating through the tubing and through the power cell back to the flask of uranium solution, where a Geiger counter records the level of radioactivity. There is a time lapse, and Guillen brings the viewers up to date: "The experiment began at high noon with the Geiger counter registering 3,760 counts per minute. As you can see," he says pointing to a meter, "after only a couple of hours, the radioactivity was cut down more than half . . . a reduction that would have taken billions of years to happen naturally." Indeed, starting with uranium-238, the end product of a long chain of natural radioactive decays is stable lead. The half-life of the entire process is about 4.5 billion years. *Every step in that process involves the emission of nuclear radiation.* Guillen seemed to be telling the audience that the Patterson cell accelerates the conversion of uranium to lead by some magical process that produces no radiation and requires no energy. That would violate everything that's known about nuclear physics. What, then, could be going on?

If a stage magician pulls a rabbit from a hat, those in the audience may not know where the rabbit came from, but unless they're hopelessly naive they know it isn't magic. It's a trick. And not a terribly difficult trick either, according to professional magicians. But how much easier it is to fool an audience with a complicated scientific apparatus. The television audience must accept on faith that the experiment is what the scientist says it is. It is as if, instead of pulling the rabbit out of the hat, the magician simply looks into the hat and assures the audience that the rabbit is there.

Could it be, for example, that the uranium ions in the solution were merely being absorbed by Patterson's beads? It would be a simple matter to see if the beads used became radioactive during the course of the experiment, but that was not done.

Guillen, however, is impressed. "The ability to neutralize radioactivity," he says, "has attracted big-name scientists like Norm Olson from Hanford, Washington, where the government stores billions of gallons of high-level waste." A head, identified as belonging to Dr. Norman Olson, says, "I'm really encouraged by what I've just seen. The plan now would be to take it back to the Hanford labs, test it out under controlled circumstances, and fully prove it as far as we can go with it."

I had never heard of this "big-name scientist," so I went to the standard science directories. He was not in any of them. I finally tracked him down at Battelle's Pacific Northwest Laboratories in Hanford. He is not a research scientist; he's a program director in the Environmental Technology Program Office. He turned out to be a pleasant fellow who was genuinely optimistic about the potential of the Patterson cell to eliminate nuclear waste but way over his head in nuclear physics. He admitted that physicist colleagues had warned him the Patterson cell was a scam, but he said he was keeping an open mind. I asked Olson how the neutralization worked. The Patterson cell, he explained, ordinarily makes it possible for hydrogen or deuterium nuclei to fuse with one another, generating heat, but if there are radioactive nuclei present, the hydrogen or deuterium nuclei can fuse with them, transmuting them to stable isotopes.

I suppose you have to admire people with an open mind, but a nuclear physicist would have more trouble swallowing the transmutation claim than the original cold fusion claim of Pons and Fleischmann. *Good Morning America* used Olson's comment to make it appear that the CETI claim is an important story. CETI, in turn, used Olson's appearance on *Good Morning America* over and over in its advertising, where it was cited as proof that the Patterson cell is taken seriously by "big-name scientists." Nothing more has been heard of Dr. Olson's test of the Patterson cell.

There are parallels with the CBS "update" on Joe Newman and the Energy Machine: Patterson is no doubt good for ratings, and so is Joe Newman. Both stories used a sound bite from someone purported to be an expert to support the claim; neither story had even a token critic. Both correspondents began by expressing a measure of skepticism but after personally witnessing a demonstration, recorded by the camera, declared the test to be a success. The audience was given no reason to doubt that a miracle had occurred. In spite of these similarities, however, the two stories were designed to appeal to people with very different worldviews.

In the CBS treatment of Joe Newman's Energy Machine, much was made of Newman's limited education and backwoods roots. No white lab coats here; this is a man with grease under his fingernails—not a scientist but a mechanic. He is vouched for by practical people, not college professors, and the correspondent readily acknowledges his own limited science background. It's the little man, fighting both the scientific establishment and government bureaucrats. The Joe Newman story was designed to appeal to those who want to see arrogant authority taken down a notch.

Contrast that with *Good Morning America*'s references to Guillen's scientific credentials, which are exceptional for a television correspondent. Even so, ABC exaggerated them; Charles Gibson introduced him as a "professor of physics at Harvard"—a disturbing stretch of the truth. Norm Olson was described as a "big-name scientist," and Patterson himself was said to have a "distinguished track record." Patterson even wore a white lab coat and carried a clipboard. All the symbols of scientific authority were paraded before the audience in an attempt to portray the Patterson cell as mainstream science. The story was designed to appeal to those who revere authority.

But what of James Patterson? Has it dawned on him yet that his beads are not actually generating energy? Has he, like Joe Newman, crossed the line from foolishness to fraud?

THE FORNICATING PRIESTS

Scientists and inventors rarely set out to commit fraud. I see no reason to doubt that, at least in the beginning, James Patterson, like hundreds of others before him, including Joe Newman, Pons and Fleischmann, Garabed Giragossian, and Robert Fludd, believed he had discovered a source of infinite, free energy. At some point it must become obvious to Patterson, if it hasn't already, that his cell simply doesn't do what he imagined it did.

What of Pons and Fleischmann? We left their story unresolved. When we left them in chapter 5, they had just tumbled from the heights of their triumphant reception by the House Science Committee to the roasting they got in absentia at the American Physical Society meeting in Baltimore. Nevertheless, Pons and Fleischmann both showed up a week later for the annual meeting of the Electrochemical Society in Los Angeles, which had organized a special cold fusion session of its own. By now, every major scientific conference seemed to turn into a referendum on cold fusion. The Utah chemists had been promised Los Angeles would be a friendlier venue. The field of electrochemistry was something of a backwater before cold fusion came along, and the electrochemists seemed determined not to let anyone puncture their balloon; they issued a call for papers that "verified cold fusion." Surely, I thought, the organizers did not mean to suggest that papers contradicting cold fusion would be rejected? Such an obvious affront to the principle of open scientific exchange, it seemed, must be an error or a prank. But a call to the conference chairman produced the response: "Since the subject of the session is fusion, papers that don't report fusion would not be appropriate."

Even in this friendly environment, however, Pons and Fleischmann were pressed to respond to the criticisms voiced in Baltimore a week earlier. They took a somewhat conciliatory position, admitting that their initial claim to have detected neutron emission was based in part on faulty evidence. They promised to repeat the measurements using more sensitive techniques. Nevertheless, they expressed confidence that it was indeed fusion that they were observing. "What else could it be?" Fleischmann asked, repeating the claim that the heat produced was much too great to be due to chemical reactions. It must be, he said, that two deuterium atoms fuse to form helium-4 by some previously unknown process that generates heat but little or no nuclear radiation. Nevertheless, he said, they were undertaking a new test, free of the flaws that had marred their previous work. The results, he promised, would persuade the most cynical skeptics of the validity of their breakthrough.

By that time, a number of scientists had pointed out that if fusion had taken place as Pons and Fleischmann claimed, the decisive evidence must already exist. The helium that is the final product of deuterium fusion should still be trapped in the metal lattice of their palladium cathodes. It would be a simple matter to analyze the cathodes for its presence. Several government laboratories offered to conduct the assay. All they would need would be a small piece of one of the used cathodes; it would take no more than a day or so to do the analysis. Pons and Fleischmann agreed that it would be the definitive test, but they refused all offers of help, insisting they were obligated to have Johnson-Mathey, the British company that supplied their palladium, perform the analysis.

Meanwhile, nearly a thousand scientists from every corner of the globe were preparing to gather in Santa Fe, New Mexico, on May 23–25 for what was billed as "the cold fusion shootout." Tens of thousands of others would gather to watch the proceedings live via satellite. The secretary of energy, Admiral James Watkins, had asked a distinguished Nobel laureate, J. Robert Schrieffer, to organize the Santa Fe meeting and to ensure that all points of view were fairly represented. The objective was to exchange every scrap of theoretical and experimental evidence that might shed light on the question. Schrieffer invited Pons and Fleischmann to lead off the conference, and they accepted, promising to present the results of their new tests. But that was before the raccoon attack.

Shortly before the Santa Fe conference was set to begin, word came from James Brophy, the vice president for research at the University of Utah, that there had been an unfortunate accident. In preparation for the definitive experiment, deuterium had been loaded into the palladium cathode of the electrolytic cell, a process that was said to require a period of several days. Unfortunately, just as they were preparing to carry out the test, a raccoon wandered into a transformer at the university. It was not only disastrous for the raccoon, it temporarily interrupted power to the building that housed Pons's laboratory. During the outage, the deuterium diffused back out of the palladium cathode, forcing the chemists to start all over again. The Santa Fe meeting would have to go on without Pons and Fleischmann.

Moshe Gai, on the other hand, could not have been kept away from Sante Fe. He was not greatly disappointed that Pons and Fleischmann would not be present; as far as he was concerned, they had already been proven wrong. But there were by that time several groups around the world claiming to detect neutrons at levels millions of times lower than those reported by Pons and Fleischmann, including a group at Los Alamos National Laboratory, a few miles from Sante Fe, two groups in Italy, and a group led by Steven Jones at Brigham Young University. With Pons and Fleischmann ducking the Sante Fe conference, and little in the way of support for their claims, the question of whether there might be an effect at extremely low levels took on added significance. At such low levels, fusion would have no practical value, but it would be important scientifically. It would also be difficult to prove one way or the other.

Following the meeting in Baltimore, Moshe Gai and Kelvin Lynn had focused on increasing the volume of material in their cells to achieve even greater sensitivity. They wanted to determine if there was neutron emission at *any* level. When it was Gai's turn to deliver the results of the Yale-Brookhaven collaboration, he explained that they now had all the sensitivity they needed to unambiguously test even these very low-level claims, and they still found nothing. The other groups, Gai said bluntly, were simply being fooled by statistical noise in the cosmic ray background.

The Sante Fe meeting provided no support for the claims of heat production by Pons and Fleischmann, and their failure to show up was taken by most scientists as further evidence that the whole cold fusion episode was about over. In Salt Lake City, however, Pons and Fleischmann issued a statement on the news coming out of Sante Fe, declaring that their critics would be forced to "eat a lot of crow." Their experiment was going better than ever, they insisted, now producing a hundred times more energy in the form of heat than it used in electricity.

The unambiguous test of whether there had been fusion, however, as Pons and Fleischmann had acknowledged in Los Angeles, was the presence of helium in the palladium cathodes. Perhaps, as they suggested, deuterium nuclei in a palladium lattice prefer to fuse by some hitherto undiscovered process in which neither neutrons nor gamma rays are emitted and the excess energy goes directly into heat—but helium would still be produced. As the days passed, however, no word came out of Utah about the promised helium assay. Pons and Fleischmann were still not returning calls from other scientists—certainly not from scientists who publicly questioned their claims—so I called James Brophy, the vice president for research. Brophy was friendly and enthusiastic but caught in a difficult position. Although he firmly believed in cold fusion, he was clearly uncomfortable about the way Pons and Fleischmann had conducted themselves. He assured me that Johnson-Mathey would complete the helium analysis soon.

After that, I made it a habit to call Brophy every day or so to see if there was anything new. He seemed to enjoy our conversations and would excitedly tell me about some new confirmation, perhaps in the Soviet Union. The next time I would call, he would explain that the Soviet report had been withdrawn—but there was an even better result from a group in Italy. However, there was no helium analysis yet. Finally, on June 2, it was Brophy who called me. Clearly excited, he said a representative from Johnson-Mathey was bringing the results of the helium assay to Salt Lake City, and the university had scheduled a press conference for June 6 to announce the results.

I called the University of Utah public information office on June 6 and asked them to fax the press release to me. There was no press release, I was told; the press conference had been canceled. I got through to Jim Brophy and asked what had happened. His voice trembled. Professor Pons, he said, had decided not to release the analysis. "Why?" I asked. Clearly shaken, Brophy replied in a barely audible voice: "He said it wouldn't be proper because it hasn't been peer-reviewed." Peer-reviewed! In March these two had trumpeted their claim to the world without peer review. "They will publish the results of the helium assay later in the summer," Brophy rasped. I knew, and Brophy knew, the results would never be published.

Consciously or unconsciously, the strategy of true believers is to isolate themselves from skeptics. Pons and Fleischmann had done so from the beginning with a curtain of secrecy. Several opportunities had arisen in the first weeks when they might have spared themselves greater embarrassment by acknowledging the possibility of error and inviting colleagues to examine their results. Each time, they chose instead to erect a higher fence, bolstering each other's courage like two schoolboys planning mischief.

They had first persuaded themselves that deuterium nuclei could be fused by squeezing them together in a palladium cathode; they had then convinced themselves that the fusion generated heat but no radiation; perhaps they had even come to believe that ordinary hydrogen could somehow fuse—it never pays to underestimate the human capacity for self-deception—but it's difficult to see how Pons and Fleischmann could have fooled themselves any longer once they had the results of the helium assay. They must have known at that point that there had been no fusion.

They were trapped by the very enormity of their claim. They could not now acknowledge that there had been no fusion without admitting that they had previously exaggerated or fabricated their evidence. They had become world figures, so now stood to be disgraced before the entire world. What began as wishful interpretations of sloppy and incomplete experiments had evolved into deliberate obfuscation and suppression of data. On June 6, 1989, just seventy-five days after the Salt Lake City announcement, cold fusion had clearly crossed the line from foolishness to fraud.

In July the Department of Energy panel headed by John Huizenga submitted its preliminary report stating that additional research into cold fusion was not justified. In the following weeks, Fleischmann returned to his home in rural Tisbury, England, for undisclosed medical treatment. Stanley Pons, after resigning his position at the University of Utah, disappeared for a time, then resurfaced in Nice, living the good life in the south of France. He had been hired to work on cold fusion by Technova, a subsidiary of Toyota. Martin Fleischmann later joined him there. Cold fusion had been officially declared dead, but was there a faint pulse?

Many of the scientists around the world who had followed Pons and Fleischmann into the quagmire of cold fusion now joined them in insisting that the effect was real and in promising to reveal new information that would prove it beyond any doubt. Three Italian physicists from the Institute of Physics in Milan, Giuliano Preparata, Tullio Bressani, and Emilio Del Guidice, were among those who claimed to have proof of cold fusion. Giovanni Pacci, the science editor of the influential Italian newspaper *La Repubblica*, described the three Italian scientists, along with Stanley Pons and Martin Fleischmann, as "scientific frauds." He compared them to "fornicating priests" for their betrayal of science. They had, he wrote, defiled "the temple of truth."

In 1992, Pons and Fleischmann filed suit against Pacci and *La Repubblica* for $5 million, charging that the paper had libeled them. They were later joined in the suit by the three physicists from Milan. The newspaper and its science editor defiantly refused to back down. Douglas Morrison, the high-energy physicist from CERN who had closely followed the entire cold fusion episode, agreed to serve the paper as its scientific advisor in the suit.

In 1996 the Italian court issued a stinging fourteen-page decision rejecting the complaint of the five scientists and ordering them to pay the expenses of *La Repubblica*. The court noted several cases of clear misrepresentation in the cold fusion claims and observed that nothing had changed in the seven years that had passed. The evidence for cold fusion was as flimsy as ever—and Martin Fleischmann still brewed his tea on a hotplate. Stanley Pons and Martin Fleischmann, the judge concluded in the unkindest cut of all, were "separated from reality."

March 23, 1999, the tenth anniversary of the University of Utah cold fusion announcement, fell during the Centennial Meeting of the American Physical Society. More than eleven thousand physicists from around the world had gathered in Atlanta to celebrate the incredible achievements of physics in the twentieth century. In keeping with its tradition, the society allowed sessions on any area of physics. One of the sessions was organized by the dwindling band of cold fusion true believers. Only six papers on cold fusion were submitted, and two of the speakers did not show up. A few dozen curious physicists were there for the start of the session; most drifted away before it was over. Douglas Morrison, the British particle physicist who has kept an eye on cold fusion for the rest of us, was there, once again asking hard questions and enduring a verbal barrage from the believers.

Pons and Fleischmann were not there. Technova had finally given up on cold fusion. Stanley Pons was let go and is reportedly living in near seclusion on a farm in the south of France. In ten years, he had done little but repeat the flawed experiments that were done at the University of Utah. Martin Fleischmann is back in Southampton and is said to be in poor health. They have fallen out and no longer speak to one another. Fleischmann now tells people that cold fusion was a victim of a campaign of distortion by the oil industry.

DENNIS LEE AND THE FISHER ENGINE

The claims of infinite free energy we have discussed up until now all seem to have started innocently, however they may have ended. It has been a story of scientists and inventors who first fooled themselves, though they may later have resorted to deliberately fooling others. But any idea with such powerful popular appeal is bound to attract flimflam artists, ready to exploit the scientific naiveté of the general public.

A network newsmagazine producer in New York called me on July 11, 1997. He was going to Hackensack, New Jersey, the next day with a camera crew to cover a demonstration of a perpetual motion machine. The demonstration, he told me, had been announced in a full-page ad in the *Wall Street Journal*. Would I be willing to come along?

I looked up the issue of the *Journal*. The ad was for a company called Better World Technologies, Inc., and Dennis Lee. I had never heard of the company or Dennis Lee, but across the top of the page, in large, bold letters, the ad read:

NEVER PAY AN ELECTRIC BILL AGAIN!

This was the same promise Joe Newman made on the CBS *Evening News* thirteen years earlier. Curious, I agreed to go.

The ad promised "An Historic Event. A Free Public Demonstration of a Perpetual Motion Device." Other technical breakthroughs would also be demonstrated, it said, including ways to modify existing vehicles so they can operate without fuel or batteries at no cost and with absolutely no negative impact on the environment, an internal combustion engine that runs on water, a way to burn water and even weld steel with water, a way to get unlimited fresh water from the ocean, a product that can grow better crops without pesticides or chemical fertilizers, a means of neutralizing radioactive waste . . . and more. You might have thought that just a perpetual motion machine would be historic enough.

By the time the shuttle from Washington landed at La Guardia, I was half regretting my decision; it was a bright Saturday morning, comfortable for July, a day to be outside. The limo stopped in Manhattan to pick up the producer for the short drive across the river to Hackensack. The camera crew was waiting when we arrived. We were two hours early for a show that was scheduled to begin at noon. A crowd had already gathered in the parking lot of what appeared to be a vacant discount store. No one was being allowed in yet. The camera crew set up in the parking lot and began doing man-on-the-street interviews with the growing crowd.

Mostly white, middle-aged, middle-income Americans, they came from places like Garden City, Kansas, and Billings, Montana — places a long way from Hackensack. Perhaps half the crowd was retired—mostly married couples. A man in his thirties said he was an electrical engineer from Wisconsin, but most had no technical background. Several said they were "dealers." Better World Technologies, I learned, sells dealerships. There had been a private showing for dealers the day before, and there would be prayer meeting for them on Sunday, but these people couldn't seem to get enough and wanted to see the "great stuff" again. Some said they had paid as much as $100,000 for their "dealership." Others boasted that they had gotten in on the ground floor, buying their dealerships when they were selling for only $10,000.

Noon came, the scheduled starting time, but they still weren't letting anyone in. There was no shade in the asphalt parking lot, and the crowd was getting hot and restless. Some people had been waiting since 9:00 A.M. to be sure of a seat; they hadn't eaten, and the program was scheduled to run three hours. It seemed totally disorganized; there were no lines, just a sweating crowd around the entrance. About 12:40 P.M. Dennis Lee, president of Better World Technologies, Inc., arrived. Somewhat disheveled, with a scruffy beard and a belly hanging over his belt, he pushed his way through the crowd to the entrance. They set up a loudspeaker, and he took the microphone. In a pitiful croak—barely audible even with the volume turned all the way up—he apologized for the delay, explaining that he had awakened that morning with severe laryngitis. On top of that, there had been some mishaps in setting up the demonstrations: the perpetual motion machine had "lost its charge," and a tank had ruptured in the welding exhibit, burning the legs of some of the staff with "caustic." However, they were all dedicated to the cause of a better world, and the show would go on, but his wife would have to do the talking.

The camera crew was let in first so they could begin setting up, and I was allowed in with them. Finally the doors were opened. As the crowd streamed in, Lee was seated on a makeshift stage in full view. A "touch therapist" was methodically going over him from head to foot with her hands about three inches from his body, smoothing out his energy field. He appeared to be in a trance. No one seemed to think this was strange.

It was a full house—perhaps seven hundred. A couple of hundred more had to watch on closed circuit television. When everyone was seated, Lee, in a barely audible croak, introduced his wife. She began with a prayer, ending with "We pray that everyone knows that all this came from You." She then began to read haltingly from a handwritten script; Lee had to help her out. His croak had noticeably improved—maybe it was the touch therapy. He took the microphone himself. "My voice is coming back," he rasped. "There is a God, and he wants you to have this show." The crowd applauded. His voice continued to grow stronger. He eventually reached a full bellow, and he went on almost nonstop for the next four-plus hours.

Before the first demonstration, Lee laid down some ground rules. On this day, there would be no questions—and no one would be allowed to examine the machines. There would be a more scientific presentation at a later date, he promised. "Scientists, with their big words," could ask questions then. It was the first of many sneering references to scientists that would be scattered throughout the show. "Today there will be no Doubting Thomases. Today is for the people." Applause again.

He began by explaining that perpetual motion is not the big deal scientists make it out to be. "Gravity is always there," he said. "Isn't that a perpetual energy source? The Earth has been going around the Sun for billions of years. God must be doing it." I looked at the people around me; they were nodding their heads in agreement. I often hear some version of this misunderstanding from freshman physics students. There is a lot of energy stored in the motion of the Earth. It's like the energy stored in Garabed Giragossian's flywheel. God does not need to do any work to keep Earth going around the Sun, because there is no friction. But gravity is not a "source" of energy, as Robert Fludd discovered with his waterwheel. A machine cannot be built on Earth that runs off the Earth's gravity.

There is also energy in the air around us, Lee explained—heat energy from the Sun. You can use that energy to make an incredibly efficient heat pump, he was saying, just as the lights went out. The staff shuffled around in the dark, while the temperature, which was already high, rose into the nineties. Finally they got the lights back on. The air conditioning had overloaded the circuits, we were told. No one seemed to notice the irony of this happening while Dennis

Lee was talking about how Better World Technologies could extract energy from the air.

Over the next couple of hours, Lee and his staff demonstrated one incredible invention after another, linked only by the fact that they all seemed to violate some fundamental law of nature. In one form or another, I had seen most of these venerable hoaxes before: the automobile that runs on water; the "power controller" that makes electric motors run at 100 percent efficiency. Power plants and automobiles could also run at 100 percent efficiency, he said. "The polluters make sure they only run at thirty percent efficiency so they can sell more oil. The Energy Department is just looking out for industry. We need an Energy Department that looks out for the people." More applause.

There was a "modified" car engine that Lee said could produce far more torque than it was rated at. With the engine running at full speed, he engaged a clutch that connected the shaft to an ordinary Sears torque wrench, bringing it to an abrupt stop. The torque wrench was bent double. The audience gasped in amazement and then broke into applause. The torque rating of an engine, however, refers to the continuous, or "dynamic," torque while it continues to run at speed. A torque wrench measures static torque. Abruptly stopping the engine expended all the stored flywheel energy of the rotating crankshaft at once. It was precisely the same confusion that had led to the humiliation of poor Garabed Giragossian. Dennis Lee, though, was demonstrating his machine not for a panel of renowned scientists but for a technologically unsophisticated audience with dreams of multiplying their modest savings.

There were more modern-sounding marvels as well, such as a device for extracting electrical energy from the neutrino flux, and the "medical laser camera" that can reveal either your skeletal structure or your soft organs through three feet of concrete. The audience was only shown a fuzzy video of that one. There was even a device that neutralizes radioactivity—apparently Better World Technologies was prepared to meet the competition from the Patterson cell. "We don't trust scientists," Lee snorted. "They told us we couldn't transmute elements—well, we neutralized cobalt-60.

We won't demonstrate it for scientists again—we'll demonstrate it for the people, and the scientists will come along." Huge applause.

It was classic flimflam. Dennis Lee or one of the staff might go into great detail about how an alternating current can be rectified to produce a direct current. With the audience nodding their heads that they understood, the speaker would say something like "That's what we're doing here, we're rectifying rotary motion." There would be no pause to allow the audience to reflect on whether this analogy made any sense. Words and images kept coming rapid-fire.

The audience, already exhausted by the long wait in the sun, had to struggle to keep up. If there was a momentary break in the action as they raced from one miracle to another, Dennis Lee would plug the gap with antigovernment, anti-industry, antiscientist, pro-God harangues. No form of authority was spared. He even made references to his jail time—naturally, his incarceration had been part of a plot by the greedy polluters to suppress the technologies that might save the world. "I never took a course—I'm really not a very bright man—but I'm God's man." People stood and cheered.

Near the end of this long day, forms were passed out to register for a free electric generator, installed in your home, that will produce fifteen times as much electricity as you need, at no cost. What does the company get out of this? You agree to give them the excess electricity, which they will sell to others. "It's your chance to be among the lucky one out of fifteen who will never pay an electric bill again."

Now, don't expect to get your generator installed right away. Dennis Lee explained that if they bring out this technology too quickly it will wreck the economy. They love America too much to do that—unless the polluters try to block them. "The scientists do whatever the government tells them to do. We want what's best for America, but if it comes to war, the Fisher engine will destroy them. A war with us could be a very bad thing for the United States."

There were no breaks scheduled, but several hours had passed, and I paid a much-needed visit to the men's room. There's a sort of camaraderie among men standing before a row of urinals that makes it easy to strike up a conversation. I asked the man next to me, who looked to be in his fifties, if this was the first time he had seen the show. "Oh no," he replied, in the condescending tone of the insider talking to the uninitiated. He explained that he owned a dealership in some town near Detroit and went to all the shows to see all the new inventions that are about to come out. "What is it that dealers actually sell?" I asked. "Everything seems to be in development." There was a pause. "Oh, lots of stuff," he said uneasily. I pressed him. "Well," he finally replied, "my favorite is the Sonic Bloomer." What does that do? "It uses sound at a frequency you can't hear to make flowers and vegetables grow better. I sold one to my father-in-law, and his tomatoes are twice as big as mine." I had a hunch that this was the only thing he had ever sold.

The centerpiece of the show was to be the Fisher engine, a perpetual motion machine that draws energy from its surroundings. Dennis Lee explained it this way: you start with a working fluid that vaporizes readily at room temperature. The vaporization is used to drive a piston. As the vapor expands behind the piston, its temperature will drop. With the help of "the world's most efficient heat pump," which Lee claims to have invented, the vapor is condensed. The liquid is then returned to a reservoir to drive the next stroke of the piston. The law of conservation of energy is not violated by the Fisher engine, Lee explained in his now fully restored voice. Since the engine draws its energy from the ambient temperature, it really runs on solar power. He described this as "the most important discovery in mechanical energy in history."

Well, not quite. In the first place, the idea isn't exactly new. It was "invented" in Washington, D.C., in about 1880 by Professor John Gamgee, who called it the Zeromotor. He managed to convince the chief engineer of the U.S. Navy, B. F. Isherwood, of the feasibility of the idea. Isherwood dreamed of a fleet that never had to put into port for coal. On Isherwood's recommendation, the secretary of the navy persuaded other cabinet members and even President Garfield himself to inspect a model of the Zeromotor.

In the second place, it didn't work in 1880 either. Although the navy invested heavily in the idea, Gamgee never quite succeeded in making a working model. The French physicist Sadi Carnot had explained why it couldn't work some sixty years before Gamgee came along. Because of the resistance of the piston, the vapor cannot expand enough to cool down to the condensation point. To get the vapor back into the liquid state, it must be refrigerated. The energy required to refrigerate the vapor to the liquid state is greater than the energy the engine produces. Carnot showed that a heat engine must always work between a hot body and a cold body. The efficiency depends on the difference in temperature. The temperature difference is analogous to the distance the water falls in turning a waterwheel, which is where we began our discussion of perpetual motion in chapter 1. Thus a machine that relies on extracting energy from the ambient surroundings would have zero efficiency. In short, it would violate the Second Law of Thermodynamics. How, then, could Better World Technologies have hoped to demonstrate a working engine?

It's easy enough—if it doesn't have to keep working. Gamgee tried using ammonia as the working fluid in the Zeromotor. The Fisher motor used carbon dioxide. Carbon dioxide, unlike ammonia, exists as a liquid only at high pressure. At atmospheric pressure, frozen carbon dioxide, or "dry ice," does not melt as water ice does; it sublimes, going directly from solid dry ice to the gas phase. The reservoir for the working fluid in the Fisher engine must therefore be kept under high pressure, about a thousand pounds per square inch at room temperature. Thus a reservoir of liquid carbon dioxide represents a very large amount of stored energy— the energy it took to compress the gas. The Fisher engine, then, is simply driven by compressed carbon dioxide. As the liquid carbon dioxide boils at room temperature, it supplies gas to drive the piston, just as a boiler supplies steam for a steam engine. The Fisher engine will keep running until the liquid carbon dioxide all boils away. That's quite enough to last through any demonstration—but not perpetually.

On this day, however, the Fisher engine didn't run at all. As Dennis Lee was going through his spiel on the stage, Dr. Victor Fisher himself was behind the curtains preparing his engine for the demonstration. Suddenly, from backstage, there was a loud hiss of gas released at high pressure. A few minutes later Dr. Fisher emerged. He received warm applause from the crowd, but he brought bad news; they were unable to get the engine started. The unexpectedly large crowd, he explained, had overtaxed the air conditioning, raising the temperature in the room to 87°F., the critical temperature of carbon dioxide. They should have gotten it started early in the morning while it was still cool, he said ruefully; then it would have kept the temperature in the building down just by running. That too would be a violation of the second law of thermodynamics. Dennis Lee has broken a lot of laws, but he hasn't broken the laws of thermodynamics.

I got a call a few weeks later from the television producer. Network executives had decided not to use the Dennis Lee story in the newsmagazine. The story was too technical. Television is more comfortable with human interest stories. No wonder the public has trouble distinguishing the hucksters from the experts. There is no one to tell them which are which.

Meanwhile, Better World Technologies and Dennis Lee continue to hold perpetual motion demonstrations, but even those who own a dealership must still buy their electricity from their local power company. Delivery of the free generators hasn't started yet.

It is easy to dismiss the people who packed that stuffy makeshift auditorium in Hackensack for almost five hours as foolish, and even to feel that they deserve to be fleeced. But I came away with the impression that these people were somewhat more knowledgeable about technology than the average citizen, and mistrust of authority is not at all unreasonable; all sorts of outrageous claims are made in the name of science. Extending mistrust of scientific claims to include mistrust of the underlying laws of physics, however, is a reckless gamble. And yet, as we will see in the next section, people who have technical backgrounds and hold highly responsible positions fall into the same trap.

HONEY, I SHRUNK THE HYDROGEN

The business editor of The *Princeton Packet* was working on a story about a small, high-tech company called BlackLight Power that was moving to Princeton. A pair of major utility companies had invested $10 million in the company, and it needed a larger laboratory. She wanted to know what I could tell her about Randell Mills and the technology he says will produce an inexhaustible supply of low-cost, nonpolluting energy. Mills, the founder of BlackLight Power, said it was "the most important discovery of all time . . . up there with fire." I asked her to wait a minute while I poured a cup of coffee; this was going to take a while.

Randell Mills, M.D., is a 1986 graduate of Harvard Medical School. Bored by the practice of medicine, he joined the cold fusion stampede in the spring of 1989, attempting to replicate the Pons and Fleischmann experiment. Two years later, long after most scientists had returned to more productive lines of research, Mills held a press conference in Lancaster, Pennsylvania, to announce that he had solved the mystery of cold fusion. It was not really fusion at all, he explained; it was a catalytic process that allows hydrogen atoms to make a transition into a state below the ground state. In that state they are much smaller than normal hydrogen atoms. He calls them hydrinos. The transition to this remarkable state releases large amounts of energy.

"A state below the ground state" is a contradiction, of course. *Ground state* refers to the lowest energy state a system can have. We earlier used the analogy of an alarm clock whose spring is completely unwound. Mills claimed to have developed "a grand unified theory of classical quantum mechanics" that explains how it is possible to have a state below the ground state.

What led to all the cold fusion claims, according to Mills, was that the shrunken hydrogen atoms can come much closer together. If they are deuterium, he argued, they may even fuse from time to time, which explains the erratic reports of neutron emission. It also explains why heat is produced even with ordinary hydrogen. Ordinary water, in fact, is preferable to heavy water in Mills's view, because it avoids fusion and the accompanying nuclear radiation.

Scientists reacted to Mills's claim exactly as they had to Joe Newman's—they ignored it. In the first place, there was really nothing to react to. His "theory" reminded me of my thesis advisor's comment when I referred to my first scientific paper as a "theory." "It's a theory," he said gently, "to the extent that it was done with a pencil." Nor had Mills offered any experimental evidence for his claim.

The energy levels of atoms are studied through their atomic spectra. Atoms can be excited into higher states by radiation. As they spontaneously cascade back to the ground state, they emit photons of very precise energies, corresponding to the energy differences between levels. These are "quantum jumps" from one state to another. The process comes to a stop when the atoms reach the ground state. An element's spectrum is thus made up of a series of sharp lines—very pure colors—that are its fingerprint. Sodium, for example, has a pair of strong spectral lines in the yellow region. These two lines account for the yellow light from sodium vapor street lamps.

The explanation of the spectrum of hydrogen was one of the first great triumphs of quantum mechanics. Every line is accounted for perfectly by the theory. There is no line corresponding to a "hydrino" state. The spectrum of hydrogen can be regarded as the platform on which our entire understanding of atomic physics is built. This is because the hydrogen atom, alone among the elements, represents an exactly solvable problem. There are only a handful of exactly solvable problems in physics. Problems involving more than two bodies can only be solved by approximate means, although the approximation can be as good as you wish. The hydrogen atom is a two-body problem, consisting of a single proton and a single electron, and can therefore be solved exactly, providing enormous insight into more complex problems. There is no physical system in the universe that is better understood than the hydrogen atom.

I thought the 1991 press conference was the last anyone would hear of Randy Mills and his hydrinos. It was the same mistake I made with Joe Newman. To my surprise, I learned some months later that Mills had formed a company called HydroCatalysis. The first customer for his device, which looked a lot like a cold fusion cell, was NASA, which wanted to evaluate it as a possible way to power a spacecraft on a mission to Pluto. It was Pascal's wager, a little money invested in a project that has little chance of working but a huge payoff if it does. The results of NASA's tests were reported to be "inconclusive." That's NASA talk for "it didn't work"—but if you said "it didn't work" you'd have to explain why you'd paid all that money for an ordinary electrolysis cell.

HydroCatalysis, nevertheless, reemerged a few years later as BlackLight Power and turned to utility companies for backing.

Large corporations often set a certain amount of risk capital aside for investment in unproven technologies. It's cheap if they invest early, so they feel they can take bigger risks. It's not an unreasonable policy—as long as the odds of success are not zero. But those who bet on hydrinos are betting against the most firmly established and successful laws of physics.

Nevertheless, the two utility companies invested a total of $10 million in BlackLight Power. If it worked, it would be worth billions; if it failed they would only be out a few million. But were the odds of it working zero? The officials at the utility companies who were responsible for venture capital investments didn't think so. They mistrusted the authority of science. That's not the same as mistrusting scientists. You should mistrust scientists; all sorts of outrageous claims are made by people who represent themselves as scientists. But these companies managed to get it just backwards: they trusted Randy Mills and mistrusted the underlying laws of physics.

The woman from The *Princeton Packet* listened carefully to all this. "But is it possible," she asked, "that the laws of physics are wrong in this case, and Randell Mills is right?" It depends on what you mean by "possible." It is, in some sense "possible," I suppose, but if he is right, the foundations of modern physics, which seem to be marvelously successful, are wrong. A better way to phrase the question is "What are the odds that Randell Mills is right?" To within a very high degree of accuracy, the odds are zero. It's Pascal's wager again.

THE PODKLETNOV GRAVITY SHIELD

The problem is that we all *want* to see a miracle. And perhaps scientists more than others. They have been drawn to science by its promise of miracles. Of course, there are scientific miracles—more, it seems, each year—or at least advances that would have seemed like miracles a few years ago. Besides, who could blame venture capitalists for investing in hydrinos when there are scientists at NASA investing in gravity shields?

Among the dozens of Small Business Innovative Research Phase II awards announced by NASA in 1999, one for $600,000 went to Superconductive Components of Columbus, Ohio, to fabricate a layered, twelve-inch superconducting disk. The disk was part of an experimental test of the 1992 claim of a Russian physicist, Eugene Podkletnov, that an object placed above a spinning superconducting disk showed a decrease of about 2 percent in weight. The work had been done in Finland, using a superconducting disk fabricated in Moscow from one of the new class of high-transition-temperature ceramic superconductors.

Superconductivity itself was a miracle when it was discovered by Dutch physicist Kamerlingh Onnes in 1911. He found that mercury loses all electrical resistance when it's cooled below 4° on the Kelvin scale (−269° C.). This was totally inexplicable in terms of classical physics and was one of the first hints of the quantum revolution that was about to take place. Even so, it was another forty-two years before the great American physicist John Bardeen and two of his students, Leon Cooper and Robert Schrieffer, gave a quantum mechanical explanation of superconductivity. There was yet another miracle in 1986 when Alex Mueller and Georg Bednorz at the IBM Laboratory in Zurich discovered a new class of ceramic superconductors with much higher transition temperatures. All three discoveries earned the discoverers Nobel Prizes.

Could Podkletnov's report of a gravity shield be another miracle? There was no stampede among scientists to find out. Gravity is a weak force compared to electromagnetism. It is not a trivial task to rule out error in such measurements or to fabricate the superconducting ceramic disks that are needed. Nor did Podkletnov, who was not a well-known scientist, seem to have a very clear idea of why it should work. The paper was not published in a particularly prestigious journal, and a coauthor asked that his name be removed from the paper—not a good sign.

Besides, claims of antigravity devices seem to come up every few years, only to fade away in a matter of weeks or months. On December 26, 1989, I was called at my office by Bill Broad, a science writer for the *New York Times*. The latest issue of *Physical Review Letters*, perhaps the most prestigious physics journal in the world, had an article by two Japanese scientists who claimed to have measured a weight loss of 0.5 percent in a spinning mechanical gyroscope—but only when it spins counterclockwise. I was apparently the only physicist he could find at work on the day after Christmas. I predicted the claim would be quickly refuted, and in just six weeks a group at the Joint Institute for Laboratory Astrophysics in Boulder, Colorado, repeated the experiment with much greater sensitivity and found no effect at all. As a result of being quoted in the *New York Times* story, however, I was flooded with calls from around the world from people claiming they had discovered the antigravity effect first; one claimed to have a patent on it, and several pointed out to me that that's the way flying saucers work. Another said he had done his research with a Frisbee.

No one took Podkletnov's claim seriously enough to attempt to replicate his work—except at NASA. NASA has spent four years and more than $1 million attempting to repeat it. So far the results have been "inconclusive." In this case, that means researchers measured a weight change of only two parts in one hundred million, which they admit could have just been an artifact of the measurement. Any weight reduction at all would be a revolutionary discovery, but small effects always raise questions about flaws in the experiment.

They even brought Podkletnov to the United States to see if he could help. He said he was puzzled—it had worked for him. The measurements were done on a six-inch disk, fabricated by Superconducting Components under a $70,000 Phase I award. Maybe, he suggested, they needed a bigger disk. That's where they are now, in Phase II, fabricating a twelve-inch disk. The new disk is layered, to multiply the effect.

In an interview with the science reporter for the *Columbus Dispatch*, the associate director of the Space Science Laboratory at the Marshall Space Flight Center in Alabama sought to explain why they were continuing to pour money into a project that seemed to have so little prospect for success. "Let your imagination run wild," he said. "What could you do if you could cut gravity by fifty percent or negate gravity altogether?" The reporter called me and repeated the question. "What could you do?" Well, for one thing, I explained, you could build a perpetual motion machine. Imagine, if you will, a wheel mounted on a horizontal axis. If a shield that reduces gravity by any amount is placed under one half of the wheel, the wheel will be unbalanced, causing it to rotate—continuously. This is hardly a new idea; it was proposed about 250 years ago. All that was missing back then was the shield. It's still missing.

You can view this two ways: either you accept the first law of thermodynamics, in which case the fact that a gravity shield would let you build a perpetual motion machine becomes proof that such a shield is impossible, or you figure the First Law of Thermodynamics might be wrong and launch a search for a gravity shield. The scientists at NASA chose the second option. They are betting against the laws of thermodynamics. No one has ever won that wager.

THE BACKWOODS WIZARD

We should, of course, end our discussion of perpetual motion and infinite free energy where we began in chapter 1, with Joseph Newman and the Energy Machine. Not long after I called Newman at his home in Lucedale, he announced publicly that he was leaving Mississippi. He had found new backers and was moving west to form Newman Energy Technologies Corporation in Castle Rock, Colorado. "It is my personal belief," he explained, "that the God in which I have chosen to believe has provided me with a new direction for my technology."

God's plan, according to Newman, is to completely decentralize the production of energy. Every home, business, and farm would have its own Newman Energy Machine, producing unlimited, pollution-free energy. With this abundant energy, Newman explained, saltwater could be made fresh and the desert turned into an oasis. Moreover, this new technology would be available in time to blunt the impact of the Y2K disaster that was certain to strike on January 1, 2000, cutting off power to those still dependent on the electric power industry.

The newly relocated Newman Energy Technologies announced that a public demonstration of a prototype motor/generator operating far beyond 100 percent efficiency would be held on September 12, 1998, in Phoenix. People came from hundreds of miles to see the new machine, but although the device was displayed, it could not be made to run. The demonstration was a disaster. It had been built for Newman by a company in Pennsylvania, and

Newman later charged that the company had deliberately sabo-
taged the motor/generator. It was all part of the conspiracy to suppress his invention. None of this discouraged Newman from trying
to persuade those who attended the "demonstration" to invest in
the project. "Put one in your home," Newman was still telling
people, as he had fifteen years earlier on the CBS news, "and you'll
never have to pay another electric bill." It was a fresh start out here
in the West. Back in Lucedale, everybody knew that Joe Newman's
house was connected to Mississippi Power Corporation lines.

SEVEN
CURRENTS OF FEAR
In Which Power Lines Are Suspected of Causing Cancer

IT'S BEST TO AVOID POVERTY

PAUL BRODEUR WAS NOT THERE in 1996 when the National Academy of Sciences (NAS) released the results of an exhaustive three-year review of the possible health effects of exposure to residential electromagnetic fields. He no longer wrote for the *New Yorker;* there had been a major shake-up at the magazine in the summer of 1992, and Brodeur did not fit into the new style. It was his series of sensational *New Yorker* articles in 1989 that first aroused widespread public fear about power lines and cancer. Without Paul Brodeur there would have been no Academy review. Now retired from the environmental health wars, Brodeur would return to active duty one more time after learning of the Academy's conclusions.

The large conference room at the classic Academy building on Constitution Avenue near the Lincoln Memorial was crowded with reporters, TV cameras, and a few scientists. This was the most extensive, most current, and most prestigious study yet conducted of the huge body of scientific evidence on the relationship between power lines and cancer. The chair of the review panel, Charles Stevens, a distinguished neurobiologist with the Salk Institute, began by explaining the difficulty of trying to identify weak environmental hazards. Scientists had labored for seventeen years to evaluate the hazards of power-line fields; they had conducted epidemiological studies, laboratory research, and computational analysis. "Our committee evaluated over five hundred studies," Stevens said, "and in the end all we can say is that the evidence doesn't point to these fields as being a health risk."

There had been concern in the scientific community about the composition of the committee, which was generally viewed as packed with scientists who might have reason to prefer that the controversy not be quite resolved. The vice chair of the panel was David Savitz, a University of North Carolina epidemiologist who had staked his reputation on a link between EMF and cancer. It was his presence on the panel that had most concerned many scientists, but perhaps half of the sixteen panel members were involved in research related to the health effects of EMF. A report exonerating EMF could lead to the elimination of funding for their research. They might be inclined to decide it was better to err on the side of caution and simply call for more research, as some previous groups had done.

In any event, the unanimous conclusion of the panel was that "the current body of evidence does not show that exposure to these fields presents a human health hazard." There were reporters in the room, however, who had been writing stories about the dangers of power-line fields for years, following the lead of Paul Brodeur. For Lou Slesin, the editor of *Microwave News*, an influential newsletter devoted entirely to the EMF-health issue, the controversy was his livelihood. For these reporters to now write that it had all been a false alarm would be miraculous. They were scouring the report looking for soft spots.

Was there conclusive evidence that EMF is *not* a risk, one reporter wanted to know? It was the classic difficulty of proving a negative. If no clear link is found between prolonged exposure to power-line fields and cancer, could it be that just certain people have a natural susceptibility to EMF? Or is EMF dangerous only in combination with some other environmental factor? The number of possibilities is infinite, and each of them raises a new question that can only be answered by more research. And when that research is done, it can always be asked if a larger study, or more sensitive measurements, might yet reveal a problem at some lower level. At what point should researchers decide that the connection, if there is one, is too weak to identify, or the hazard, if a hazard exists, is too insignificant to be concerned about?

Stevens did acknowledge that there seemed to be a weak statistical association between living near power lines and childhood leukemia. "The question," he said, "is what causes that association." It could not be the fields; when the fields in homes were actually measured, studies showed, the association with cancer all but disappeared. What, then, accounts for the excess incidence of leukemia in such homes? "We just don't know," Stevens said, but he pointed out that neighborhoods with heavy concentrations of power lines are usually poor, congested, and polluted—all of which are risk factors for cancer.

It was clear that many of the reporters had trouble with the concept that there could be an association between childhood leukemia and living near power lines without the power lines being the cause. If the panel could not explain the statistical link between power lines and cancer, one reporter persisted, wouldn't a policy of "prudent avoidance" be justified? *Prudent avoidance* is a term coined by Granger Morgan of Carnegie Mellon University. Morgan had made something of a career of going about the country preaching prudent avoidance to worried parents, but it seemed to mean something different to everyone. Did it mean doing away with electric hair dryers or converting to candles? "We wouldn't know what to suggest people avoid," Stevens patiently explained. Since proximity to power lines and the incidence of childhood leukemia are both greatest in congested, low-income areas, the most prudent course might be to avoid poverty.

That evening on the ABC news, Peter Jennings summed up the report in one line: "Power lines do not cause cancer—and that's that." Some reporters, however, concluded that the door had been left open just a crack by the slight statistical link between childhood leukemia and proximity of power lines. For them, as for Brodeur, it was a matter of common sense: if children living near power lines have a higher risk of leukemia, power lines are to blame.

To understand how the power-line controversy could have been sustained for so long, on the basis of so little evidence, we must first go back to an earlier false alarm over a very different form of EMF: microwaves. Once again, the public alarm was sounded by Paul Brodeur.

MONKEY BUSINESS

Ellie Adair's children were off at college, and the colony of squirrel monkeys had become the outlet for her mothering instincts. Squirrel monkeys are New World primates with prehensile tails and large eyes set in tiny, expressive faces. They are gentle, affectionate, naturally clean animals. Adair worried if there were long periods between experiments. During such periods the monkeys tended to become listless, lost weight, and began to neglect their grooming. She believed they were bored and missed the extra attention. When the experiments started up again, they would perk up, their appetites would return, and their coats would become glossier. So she tried to see that all the monkeys got as much "work" as possible. Among researchers, she had the reputation of always having the healthiest monkey colony. She liked to think it was also the happiest.

Ellie Adair, a research fellow at the John B. Pierce Foundation Laboratory, associated with Yale University, was a leading authority on the body's temperature-regulating mechanism. Mammals and birds maintain almost constant temperature over wide variations in the air temperature or internal heat generation from exercise. The area of the brain called the hypothalamus is the control center of the temperature-regulating system. The hypothalamus senses the temperature of the blood that is pumped through it; at the slightest rise in temperature, it sends out instructions to increase sweating and respiration and to dilate the blood vessels that carry blood to the skin. The instructions are conveyed by a delicate interplay of chemical messengers and nerve stimulation.

In Adair's experiments she exposed the monkeys to microwaves, just as you would heat food in a microwave oven, and monitored their physiological response. There was no reason to believe microwaves at the levels used in the experiments harmed the monkeys. The monkeys could even be trained to control the level of microwaves themselves.

But in December of 1976, Adair got a call at the lab from a colleague. "You'd better take a look at the latest issue of the *New Yorker*," the caller said. She picked up a copy on her way home. In the understated style of the magazine at the time, there was an article with the simple title "Microwaves-I," by a staff investigative reporter named Paul Brodeur. One line in the article read: "It is known that microwaves exert a profound effect on the central nervous system of rhesus monkeys and other primates." Known to whom, she wondered?

Microwaves are electromagnetic radiation, waves of electric and magnetic fields that travel at the speed of light, differing from visible light only in the frequency at which the fields oscillate. We are constantly bathed in electromagnetic radiation, most of which is unseen and unfelt. Visible light makes up a very narrow region of the electromagnetic spectrum. At frequencies just below the visible spectrum, infrared radiation can be felt on our skin, warning us before we actually touch it that a stove is hot. Microwaves correspond to still lower frequencies. Although our senses do not respond directly to microwave radiation, microwaves are absorbed by certain molecules in the body, increasing the amplitude of their atomic vibrations. That amplitude is a measure of the body's temperature. At sufficient intensity, as in a microwave oven, the heating would begin to destroy cells, but in Adair's experiments, the heating was not nearly enough to cause cell damage.

Adair had assured herself that microwaves were harmless before she ever began her research with monkeys. The biological effects of microwaves had been studied for thirty years and were the subject of hundreds of papers in the open literature. The research began during World War II with the development of radar, when a technician walking near an experimental transmitter discovered that a chocolate bar in his pocket had melted. The army set up a program to evaluate any hazard to technicians and operators directly exposed at close range to the radiation from high-powered radar systems. Much of the research into the effects of microwaves is still supported by the Department of Defense.

Microwave ovens, originally called radar ranges, were a product of that research. At the time of Brodeur's article, radar ranges were just beginning to be marketed to the public. Manufacturers relied on the same experts that Adair had consulted for information about safety. There had been some initial concern about the effect of microwaves on the eyes, which dissipate heat less effectively than other organs. Washington columnist Jack Anderson had exploited this particular concern earlier, reporting an increase of cataracts associated with the use of radar ranges. There can be a slight "leakage" of microwaves outside the range. However, more careful studies found no effect, even at levels far higher than the leakage. As long as there was adequate screening, and interlocks to prevent someone from sticking a hand in when the range was on, there seemed to be no cause for concern.

But the same facts that had reassured Ellie Adair were seen through a very different lens by Paul Brodeur. A cold-war investigative reporter who began his journalistic career exposing dark secrets of the CIA, Brodeur had switched to exposing environmental and occupational hazards in 1968. Beginning with asbestos and moving on to microwaves, he found a niche sounding the alarm about the dangers of technology. Brodeur had no technical background. Instead, he approached environmental issues with a cold-war mind-set: Who had something to gain? And what were they covering up?

Since World War II, Brodeur warned, electromagnetic radiation from radar, television, and microwave communication had risen to one hundred million times the "natural" background level in New York City, due to radar, radio, and television. It was an alarming-sounding statistic but completely meaningless. In terms of power, this was still a totally insignificant level. "Natural" microwaves are due to so-called black-body radiation, the radiation given off by all warm objects, but at ordinary temperatures, most black-body radiation is in the infrared region of the spectrum. There is very little in the microwave region. That's one reason microwaves are so useful for radar and communications. It's a "quiet" part of the spectrum. Brodeur made no distinction between the insignificant levels of background microwave radiation from television broadcasting and the exposure you might get from standing in front of a radar transmitter. The list of health problems he connected to microwaves expanded beyond cataracts to include miscarriages, birth defects, and cancer.

The fact that most research on the biological effects of microwaves had been supported by the Department of Defense became for Brodeur evidence that the government was attempting to control information about its hazards. When industry scientists reported similar findings, he saw it as proof that the electronics industry was in collusion with the military. When academic scientists scoffed at the background-microwave hazard, they too became part of Brodeur's conspiracy theory.

If we have learned anything about the environment in recent years, it is that we cannot take warnings lightly or accept uncritically the soothing reassurances of authorities. We have seen the tobacco companies suppress their own studies of nicotine addiction and the health effects of tobacco smoke; the nuclear industry, chemical companies, drug manufacturers, car makers—all at times have engaged in coverups. The federal government has conspired with civilian contractors to withhold information about the spread of radioactive contamination around nuclear weapons production facilities. Was there any reason to expect the electronics industry and the federal government to behave in a more principled fashion with regard to microwaves?

Still, Ellie found Brodeur's conspiracy claims preposterous, and she was offended by the implication that any scientist who disagreed with him must be part of a cover-up. With so many scientists holding open meetings and freely exchanging results on microwaves, a cover-up would be impossible to sustain. But the question remained: Could there be some unrecognized interaction of microwaves with the body that causes serious health problems? And if there is, why were her monkeys, which were exposed to relatively huge doses of microwaves, in perfect health?

Help in answering such questions was close at hand. She recruited her husband Bob, a physics professor at Yale. Ellie met Bob in graduate school at the University of Wisconsin. She had decided to take her Ph.D. in both experimental psychology and physics, even though it meant taking a lot of additional math. It was in the Physics Department that she met Bob—who was very good at math. They were married in 1951. Ellie spent the next years raising a family. Not until the children were in school did she resume her own career. Meanwhile, Bob had gained recognition as one of the nation's foremost nuclear theorists.

Bob Adair had grown up in a staunchly union, blue-collar family in Fort Wayne, Indiana. His father had not been to college, but he had studied physics in high school and delighted in explaining to his precocious son the physics of the world around them. From the earliest time Bob could remember, when anyone asked him what he was going to be when he grew up, he always answered, "A mathematical physicist." Bob also loved baseball, but he had no aptitude for it, failing even to make a sandlot team. Years later, however, when he had become a renowned experimental physicist, occupying an endowed chair in the Physics Department at Yale University, he would write the definitive book on the physics of baseball.

Bob was sure Brodeur had to be wrong. The effect of all known cancer-inducing agents—ionizing radiation such as ultraviolet or X rays, chemical carcinogens such as tobacco smoke, and certain viruses—is to damage DNA. The damage consists of broken or altered chemical bonds, creating a mutant strand of DNA. Microwave photons can cause chemical bonds to stretch and bend but cannot come even close to severing the bonds. One of the great triumphs of quantum mechanics was the discovery that electromagnetic radiation interacts with matter only in discrete bundles of energy called photons. The energy of a photon is expressed mathematically as the product of a universal constant, called the Planck constant, multiplied by the frequency. Photons that have enough energy to break chemical bonds are called ionizing radiation. Whether electromagnetic radiation is ionizing is independent of the intensity, or number, of photons; it depends only on the energy of the individual photons.

Breaking a chemical bond with a photon is like throwing stones at something on the other side of a river. If you can't throw that far, it won't matter how many stones you throw. The lowest-energy photons capable of directly breaking chemical bonds are in the near-ultraviolet region of the spectrum, just beyond the region of visible light. These photons are about a million times more energetic than the microwave photons Ellie Adair was using. Breaking chemical bonds with microwaves would be like trying to throw a stone across the ocean.

Meanwhile, the *New Yorker* published "Microwaves-II," in which Brodeur focused on the strange situation at the American embassy on Tchaikovsky Street in Moscow. For reasons that were a mystery at the time, the Soviets had been beaming microwave radiation at the embassy for more than a decade. It is now known that the microwaves supplied the tiny amount of power needed to operate electronic evesdropping devices that had been concealed in the building during its construction. Brodeur, however, suspected that the microwaves were meant to addle the brains of embassy workers or induce depression. What shocked him was that the government had not warned employees of the health hazard. He noted that Ambassador Walter Stoessel had developed some mysterious blood ailment, and two former ambassadors had died of cancer. To Brodeur it seemed the microwaves must be to blame. People were exposed to microwaves and they got sick—it was the belief engine at work.

A few months later, Brodeur published a book titled *The Zapping of America*, drawn from his *New Yorker* articles. Spurred by Brodeur, environmental activists embraced microwaves as a new cause. The immediate impact was to almost destroy the budding microwave oven market, but the problem didn't stop in the kitchen. Every microwave relay tower, every air traffic control radar, was suddenly suspect. Roused to near fury by Brodeur, a citizens' group went to court to block the National Weather Service from installing a weather radar at Brookhaven National Laboratory, on the grounds that it would lead to miscarriages and cancer. Long Island, which juts out into the Atlantic, had found itself in the path of numerous killer hurricanes. The radar was meant to track such storms and provide timely warning to those in their path. But people feared the known dangers of howling wind and crashing ocean waves less than they feared the unproven hazard of silent, invisible microwaves. Scientifically, the issue was one of relative risk: a history of property damage and loss of life from storms, against an unproven hazard that most scientists believed was nonexistent. But judges are not scientists, and a federal judge ruled against the Weather Service. The Long Island weather radar was canceled.

Over the next few years, however, most of the public seemed to gradually lose its fear of microwaves. New studies failed to confirm the link to cataracts or other health effects, and people were discovering the wonderful convenience of microwave ovens; sales were beginning to rebound. Within a decade, there would be a microwave oven in almost every home in America—and no related increase in health problems.

Bob Adair, meanwhile, had begun going with Ellie to scientific conferences on the effects of microwaves. He was asked to present a physicist's perception of the problem at one of the conferences and, encouraged by the positive reaction, he wrote his work up and published it in the *Physical Review*. He relied on well-established physical principles to show that there was no known mechanism that could account for reports of health effects from low levels of microwave radiation. Unknown to the Adairs, however, events were underway in Denver that would shift the conflict to a new arena and again thrust the Adairs into conflict with Paul Brodeur.

THE CURRENT CONTROVERSY

In 1979, an unemployed epidemiologist named Nancy Wertheimer obtained the addresses of childhood leukemia patients in Denver and drove about the city looking for some common environmental factor that might be responsible. What she noticed was that many of the homes of victims seemed to be near power transformers. Could it be that the fields from the electric power distribution system were linked to leukemia? She teamed up with a physicist named Ed Leeper, who devised a "wiring code" based on the size and proximity of power lines to estimate the strength of the magnetic fields. Together they eventually produced a paper relating childhood leukemia to the fields from power lines. They concluded that children from homes with "high" magnetic fields from power lines were three times as likely to develop leukemia as children from homes with "low" fields.

Few scientists were aware of the Wertheimer-Leeper work at the time, and fewer still took it seriously. In the first place, the study was not "blind": she knew in advance which were the homes of leukemia victims. In the second place, the relative strength of the power-line fields was not actually measured but merely estimated on the basis of the size and proximity of power lines. The situation was ripe for investigator bias; the tendency to judge the wiring of victims' homes more critically would be almost unavoidable. If the result for a particular home disagreed with the researcher's expectation, for example, there would be a tendency to double-check the result and see if something had been missed the first time. To the researchers, it may only seem that they are being careful, but unless all the homes are double-checked, it introduces a powerful bias. The numbers, after all, are very small—childhood leukemia is a rare disease—and the shift of only a few victims' homes from "low field" to "high field" is sufficient to change the conclusion.

Scientists must constantly be on guard against this sort of self-deception. Unless studies are carefully designed to avoid it, the biases of the epidemiologist have a way of creeping into the results. To minimize the opportunity for bias, scientists rely on double-blind studies. An independent researcher might be given a list including both the homes of victims of childhood leukemia and an equal number of addresses of nonvictim children matched in age, gender, race, family income, etc., but without any indication of which are which. Without knowing which were the homes of victims and which were "controls," the researcher would rate them by whatever criteria were used to estimate the field strength. Someone else would then apply the key after the judgments were made.

But even if the study had been double blind, a "risk ratio" of only three for a rare disease such as childhood leukemia would be regarded by many epidemiologists as barely credible. The risk ratio for lung cancer from smoking, for example, is well over thirty—that is, a 3,000 percent increase in the incidence of lung cancer among smokers. Yet it took years of checking and rechecking the figures, as well as a highly plausible mechanism in terms of known carcinogens in tobacco smoke and, finally, confirming laboratory studies on animals before the cancer link was firmly nailed down.

In spite of its obvious flaws, the Wertheimer-Leeper report could not be dismissed. We are all exposed to EMF every day of our lives, and even the most slender link to cancer would be a cause for concern. There were soon reports that electrical workers suffered high cancer rates; women using electric blankets or working at computer terminals were said to suffer frequent miscarriages; suicides were reported to be occurring at an alarming rate among people living under power lines; farmers with fields crossed by power lines claimed that their cows stopped giving milk and chickens stopped laying eggs. Although none of these stories were backed up by reliable statistical evidence, each new anecdote added to the sense that something was going on.

The Wertheimer-Leeper study was soon followed by the usual "confirmations," most of which were as seriously flawed as the Wertheimer work. We saw in chapter 1, in the case of cold fusion, that important new claims tend to attract followers who see what they expect to see. Evidence that would seem much too weak to stand on its own is taken seriously if it seems to agree with what others are reporting. There was one "confirming" study, however, that had to be taken more seriously. In 1988 David Savitz of the University of North Carolina, a highly respected epidemiologist, set out to check the Wertheimer-Leeper results, using the same "wiring code" method of estimating the 60 Hz magnetic field. He also found an increased risk of leukemia among Denver children living in homes with "high field" wiring. The very important difference was that Savitz had used accepted double-blind methods. Although the increased risk was only about half as great as that reported by Wertheimer and Leeper, Savitz thought further study was clearly called for. Most scientists, however, remained highly skeptical of the purported EMF-cancer connection. Microwaves, as we saw, can induce heating. At a mere 60 Hz, however, there is not even that.

To understand the power-line controversy, we need another brief physics lesson. At such low frequencies, it's no longer meaningful to think in terms of radiation and photons. What is measured are separate electric and magnetic fields. Perhaps the greatest achievement of nineteenth-century physics was Michael Faraday's discovery in 1831 of the relationship between electric and magnetic fields: moving electric charges, such as an electrical current flowing through a wire, generate a magnetic field. Conversely, a moving or changing magnetic field will induce a current in a stationary conductor. A power line is surrounded by both an electric and a magnetic field. The strength of the electric field depends only on the voltage; the strength of the magnetic field, only on the current. Both the electric and magnetic fields surrounding a conductor drop off rapidly with distance.

Everyone agreed that the electric fields from power lines do not represent a health hazard. Because human tissue, including skin, conducts electricity rather well, the outermost epithelial layers of the skin act as a shield, preventing electric fields from penetrating into the body. The concern is with the magnetic fields, which penetrate the body—in fact, most materials—almost unimpeded. The public tended to be most concerned about high-voltage lines. Strung on gigantic towers that resemble a file of mechanical monsters marching across the countryside, high-tension lines look threatening, but the whole purpose of the high voltage is to transport power with as little current as possible. High-voltage lines therefore minimize the magnetic field.

Humans, of course, have always been exposed to a magnetic field. Electrical currents circulating in Earth's molten core act like a huge dynamo, turning Earth itself into a giant magnet. But today, we are also exposed to man-made electromagnetic fields, generated by the electrical wiring that is ubiquitous in modern society. Over the past fifty years, the per capita consumption of electric power in the industrialized world has increased twenty-fold, and our exposure to the electromagnetic fields generated by power lines and appliances has increased by a similar amount.

In spite of the enormous growth in consumption of electricity, however, in most homes and workplaces magnetic fields produced by electric power are still only about 1 percent as strong as Earth's natural magnetic field. There is one difference: electric power is supplied as alternating current. In the United States, the frequency at which the current reverses direction is 60 Hz or sixty times each second; in Europe, it is 50 Hz. Thus, as a result of Faraday's law, an alternating magnetic field interacts with our bodies in a way that the relatively constant magnetic field of the Earth does not. The result is to induce weak electrical currents in the body. It is prudent to ask if these currents affect our health in any way. Could they somehow interfere with the body's cancer defenses? If that were the case, power-line fields might not cause cancer, but might influence the growth of cancers caused by something else.

In June of 1989, The *New Yorker* carried a new three-part series of highly sensational articles by Paul Brodeur, this time on the hazards of power-line fields. The articles drew heavily from his earlier attacks on microwaves. Indeed, he seemed to draw no clear distinction between 60 hertz and 100 megahertz, which is typical microwaves—it was all just EMF. The series reached an affluent, educated, environmentally concerned audience. Suddenly, Brodeur was everywhere: the *Today* show on NBC, *Nightline* on ABC, *This Morning* on CBS, and, of course, *Larry King Live* on CNN. In the fall, Brodeur published the *New Yorker* series as a book with the lurid title *Currents of Death*. A new generation of environmental activists, led by mothers who feared for their children's lives, demanded government action.

I was asked by the "Book World" section of the *Washington Post* to review *Currents of Death*. The book was frightening all right. Brodeur was a skilled writer, and he used all his skill to build a case against EMF. His approach was taken right out of *The Zapping of America*. He described power-line fields as the most pervasive— and covered up—health hazard facing Americans. The overwhelming consensus among scientists, that no hazard existed, was for Brodeur evidence of a massive cover-up, this time involving the utilities, the government, and the scientific community. Once again he related frightening anecdotes of suffering and death from cancer. It was easy to connect the suffering to EMF; power lines are everywhere. It was the belief engine at work: people are exposed to EMF, and people get cancer. I pointed out in my review that life expectancy in the United States had doubled in the past hundred years—and most of that increase had come since the advent of electricity.

Feeling the pressure from power-line activists, the Environmental Protection Agency (EPA) convened an internal panel to set "safe limits" on exposure to EMF. Setting safe limits on exposure to any environmental factor—radiation, chemicals, particulates—would seem to be easy enough: just set the limits so low that a harmful effect is inconceivable. But any exposure above the limit is then, by definition, "unsafe." Not only does an unrealistic exposure limit impose a financial burden on society, without making anyone safer, it may actually create new risks, since we are frequently compelled to trade off one risk for another.

Already, parents were insisting that EMF levels in schools be measured. When those levels turned out to be higher than numbers Brodeur's book had described as "unsafe," they elected to transfer their children to more distant schools where the levels were lower. They exchanged the imaginary risk of EMF for the very real risk of longer travel.

In May of 1990, a preliminary draft of the EPA report was leaked to CBS News. Dan Rather, quoting from the draft, informed his viewers that the panel had found 60 Hz magnetic fields to be a "probable but not proven cause of cancer in humans." The report caused panic among viewers who had never heard of Paul Brodeur. If it was true, the dimensions of the problem were staggering. We are bathed in such fields. There is no escape. They penetrate the walls of our homes as easily as they penetrate our bodies.

The draft, however, was in error. A corrected draft changed *probable* to the far weaker *possible,* but the damage had been done. CBS never bothered to correct its news broadcast. Predictably, Paul Brodeur declared the downgrading from "probable" to "possible" to be further evidence of a conspiracy to cover up the terrible truth; others objected that even "possible" went too far and would arouse concerns that were not justified by the evidence.

At that point, Allan Bromley, who was now White House science advisor, intervened, insisting that experts from outside the EPA review the report. The external review panel, composed of epidemiologists, medical scientists, engineers, and physicists, found that the EPA report had "serious deficiencies" and recommended it be completely rewritten and then re-reviewed. Brodeur was incensed.

In late 1990 he began a series of sequels for the *New Yorker* that relied on selected anecdotes to create a menacing atmosphere of silent, invisible fields invading homes and schools—and a conspiracy to hide the truth from the public. He went into great detail about all the sickness suffered by folks living on Meadow Street in Guilford, Connecticut. They were coming down with everything from brain cancer to Osgood's knee—and there was an electric substation on Meadow Street. Such anecdotes appeal directly to the belief engine and have a powerful emotional impact. For every Meadow Street, however, there may be a Forest Street somewhere, also with a substation, where no one seems to get sick—but Brodeur wasn't interested in Forest Street. By including only data from isolated cases that supported his belief, Brodeur was committing the same error that Irving Langmuir had discovered in the ESP studies of J. B. Rhine, discussed in chapter 2. Brodeur's focus on cancer clusters is called the Texas sharpshooter fallacy by statisticians. The sharpshooter empties his revolver into the side of a barn—and then walks over and draws a bull's eye. If you're going to argue from statistics, you must use all the statistics. Few of us are statisticians, however, and the story of Meadow Street in Guilford, Connecticut, was convincing to many readers.

This latest *New Yorker* series was also turned into a book. In *The Great Power-Line Cover-up*, Brodeur fumed that the delay in issuing the EPA report meant that "Thousands of unsuspecting children and adults will be stricken with cancer, and many of them will die unnecessarily early deaths, as a result of their exposure to power-line magnetic fields." And the person responsible was Allan Bromley. He not only leveled this shocking accusation at Bromley, he charged that Bromley had acted on my advice.

I had by that time written op-ed articles for *Newsday* and for the *New York Times* cautioning that there was scant evidence connecting EMF to cancer and advising readers to await the results of the four-year study that had just been undertaken by the National Cancer Institute. It was to be the largest and most thorough epidemiological study of the EMF-cancer connection ever attempted. Even using the worst-case numbers from published studies, it was clear that EMF could not be a very significant factor in the incidence of cancer and probably was not a factor at all. There was, I argued, little risk in waiting for a more definitive answer.

Bromley was deeply stung by Brodeur's accusation, but he was not intimidated. He commissioned Oak Ridge Associated Universities, a group of research universities with no stake in the outcome, to carry out a thorough review of all the scientific information on the subject—some five hundred technical papers. The study took two years, and the panel concluded that it was not possible to establish safe exposure levels since "no hazard has been demonstrated."

The report did little to allay public fears. There was even a 1992 movie, *The Distinguished Gentleman,* starring Eddie Murphy as a petty con man who gets elected to Congress. He is transformed into a crusading environmentalist after a chance meeting with an eight-year-old who got cancer from a power line that runs by the playground. He proceeds to battle the greedy power companies from his seat on a fictitious Power and Industry Committee.

By now, however, results were beginning to come in from larger and more sophisticated epidemiological studies—and the EMF-cancer connection was getting weaker. In particular, the newer studies actually measured field strength in homes or workplaces rather than relying on estimates using some sort of wiring code. It is a general rule in epidemiology that if a better measure of a suspected agent results in a lower risk, there is almost certainly an unidentified "confounding factor."

A study of rapid weight loss, for example, might show a correlation with premature death. Does this mean weight loss programs are unsafe? Not necessarily. It may just mean that the researcher has failed to account for the fact that many fatal diseases cause the body to waste. In this case, chronic disease would be a "confounder." "Bias and confounders are the plague upon the house of epidemiology," according to Philip Cole, chair of the Department of Epidemiology at the University of Alabama.

In 1994 a four-year study of 223,000 Canadian and French electrical workers was completed. It was the largest and most sophisticated study yet conducted. The study found no overall increase in cancer risk associated with occupational exposure to EMF. Of the thirty types of cancer included in the study, only one, a rare form of leukemia, showed an increased risk, and that was based on only five cases. The director of the study, Giles Theriault, expressed surprise at the low numbers. "I don't think we have the right agent," he said.

A year later, a very similar but even larger study of U.S. electrical workers was released. The study examined the same thirty types of cancer included in the Canadian/French study. However, the American study found no increased risk for any form of leukemia, but it did find a slightly elevated risk of one rare form of brain cancer. The head of the U.S. study, David Savitz of the University of North Carolina, called for an even larger study to resolve the difference. It was Savitz, you will recall, who had repeated Nancy Wertheimer's 1979 study of childhood leukemia.

In a study breaking cancer down into thirty types, one or two false positive findings would be just about what you would expect if there was no link at all. The reason is that, by general agreement, a statistically significant finding is defined as anything above the 95 percent confidence level. Using the 95 percent standard, you might expect a false positive about one time in twenty. If the convention were, say, 97 percent rather than 95 percent, there would have been no positive findings in either study.

In fact, both studies found the cancer rate among electrical workers to be lower than for the general population. In the Savitz study, for example, the cancer rate among electrical workers was just 86 percent that of the rest of the population. This, epidemiologists explain, is merely the "healthy worker syndrome." For a variety of reasons, people with good jobs tend to be healthier—and have fewer cancers—than people who don't; they may have a better diet, more frequent medical checkups, live in less polluted neighborhoods, experience less stress, etc. These are confounding factors.

To avoid being "misled" by the healthy worker syndrome, the incidence of cancer among electrical workers with "low" levels of EMF exposure was compared to the incidence among workers with "high" exposure levels. This, however, leaves the question of where to draw the line between "low" and "high." If the statistics are poor—that is, if the number of cases of a particular cancer is small—choosing a different boundary between "low" and "high" can reverse the outcome. Epidemiology was dredging for results in the statistical noise.

By the spring of 1995, the American Physical Society had completed its own review of the EMF literature. Scientific societies are normally reluctant to give the appearance of deciding scientific truth, feeling that their job is to provide a forum for the exchange of scientific results and ideas. In the case of EMF, however, it was felt that information coming from outside the scientific community, Paul Brodeur and *Microwave News* in particular, had given the public a seriously distorted view of the scientific facts. A statement released by the APS concluded that "conjectures relating cancer to power-line fields have not been scientifically substantiated." It was the strongest position on the EMF issue taken by a scientific society.

SLAMMING THE DOOR SHUT

By that time, sixteen years had passed since Nancy Wertheimer took her historic drive around Denver. An entire industry had grown up around the power-line controversy. Armies of epidemiologists conducted ever larger studies; activists organized campaigns to relocate power lines away from schools; the courts were clogged with damage suits; a half dozen newsletters were devoted to reporting on EMF; a brisk business had developed in measuring 60 Hz magnetic fields in homes and workplaces; fraudulent devices of every sort were being marketed to protect against EMF; and, of course, Paul Brodeur's books were selling well.

Scientists, one might argue, were also thriving. Federal agencies had responded to the public alarm by funding more and more research into the interaction of electromagnetic fields with living organisms. The Bioelectromagnetic Society, formed the year before Nancy Wertheimer published her notorious leukemia study, had grown to more than six hundred members, largely on the basis of the power-line controversy.

It was into this climate that the Stevens Report was released by the National Academy of Sciences in 1996 with its unanimous conclusion that "the current body of evidence does not show that exposure to these fields presents a human health hazard." For Brodeur, retired and living in southern California, it was just more of the conspiracy. He bitterly attacked both the report and the National Academy of Sciences in *Secrets*, a book recounting his thirty years at the *New Yorker*. Brodeur seemed not to understand confounding factors; if children living near power lines have a greater incidence of leukemia, it seemed to him that power lines must be to blame.

Ironically, Brodeur's memoirs had no sooner reached the bookstores than, on July 2, 1997, the National Cancer Institute (NCI) finally announced the results of its exhaustive epidemiological study, "Residential Exposure to Magnetic Fields and Acute Lymphoblastic Leukemia in Children." In contrast to the National Academy study, which had surveyed the entire body of literature dealing with possible health effects associated with magnetic fields, the NCI study focused on the question that started it all: are power line fields associated with childhood leukemia? The NCI study would answer the question, not by reviewing the existing literature, but by undertaking its own epidemiological investigation. And it would do so on such a scale and with such thoroughness that the results would not be subject to challenge.

What was to have been a four-year study ended up taking more than seven. It was the most unimpeachable epidemiological study of the connection between power lines and cancer yet undertaken. Every conceivable source of investigator bias was eliminated. There were 638 children under age fifteen with acute lymphoblastic leukemia enrolled in the study along with 620 carefully matched controls, ensuring reliable statistics. All measurements were double blind and included the magnetic fields in the children's bedrooms and other locations in and around their homes. Each home was also assigned a wire code based on the distance and configuration of power lines.

If the National Academy of Sciences report a year earlier left the door open a crack, it was slammed shut by the NCI study. It concluded that any link between acute lymphoblastic leukemia in children and magnetic fields is too weak to detect or to be concerned about. But the most surprising result had to do with the proximity of power lines to the homes of leukemia victims: the study found no association at all. The supposed association between proximity to power lines and childhood leukemia, which had kept the controversy alive all these years, was spurious — just an artifact of the statistical analysis. As is so often the case with voodoo science, with every improved study the effect had gotten smaller. Now, after eighteen years, it was gone completely.

The NCI study was published in the prestigious *New England Journal of Medicine*. An accompanying editorial concluded:

> It is sad that hundreds of millions of dollars have gone into studies that never had much promise of finding a way to prevent the tragedy of cancer in children. The many inconclusive and inconsistent studies have generated worry and fear and have given peace of mind to no one. The eighteen years of research have produced considerable paranoia, but little insight and no prevention. It's time to stop wasting our resources. We should redirect them to research that will be able to discover the true biologic causes of the leukemic clones that threaten the lives of children.

Research funds were redirected to other priorities. The Department of Energy closed down the EMF Research and Public Information Dissemination (RAPID) Program, created by Congress in 1992. It would no longer be needed. Now, surely, the false trail had petered out.

One year later, however, an international "working group" of experts who had been involved in the EMF issue assembled in Bethesda, Maryland. Many of those on the panel had staked their reputations on a link between power-line fields and cancer, and were working on projects whose funding depended on continued public concern. Some—like Lou Slesin, the editor of *Microwave News*, whose livelihood was directly linked to the controversy— were not even scientists. The working group treated the NCI study of childhood leukemia as just one more study. It had not, after all, been replicated. After ten days of deliberation, the group issued a call for more research. By a vote of nineteen to nine, EMF was classified as "a possible carcinogen."

That depends, of course, on what you mean by *possible*. Richard Wilson, a Harvard physicist who had researched the problem, illustrated *possible* this way: Suppose someone tells you a dog is running down the center of Fifth Avenue. You might think it unusual, but it's certainly possible, and you would have no reason to doubt the story. If the claim is that it's a lion running down Fifth Avenue, it's still possible, but you would probably want some sort of supporting evidence—perhaps a report of a lion escaping from the Bronx Zoo. But if someone tells you a stegosaurus is running down Fifth Avenue, you would assume that he's mistaken. In some sense it might be "possible" that he's seen a stegosaurus, but it's far more likely that he saw a dog and thought it was a stegosaurus. Indeed, most reasonable people would agree that the possibility that there could really be a stegosaurus running down Fifth Avenue is too small to even bother checking out. Wilson concluded that the EMF working group saw a possible stegosaurus—not a possible dog or even a possible lion.

On May 1, 1999, results of a long-awaited Canadian epidemiological study of childhood leukemia were released. The massive study, covering five provinces of Canada, closely matched the NCI study in the United States. The Canadian study found no relationship between exposure to residential electromagnetic fields and leukemia in children.

By now, the total cost of the power-line scare, including relocating power lines and loss of property values, was estimated by the White House Science Office to be in excess of $25 billion. In all that time, however, there had not been a single successful lawsuit based on health effects from electromagnetic fields. In the next chapter we will learn why.

EIGHT
JUDGMENT DAY
In Which the Courts Confront "Junk Science"

ELECTROMAGNETIC FIELDS ATTRACT SHARKS

FOUR-YEAR-OLD MALLORY was suffering from Wilms' tumor, a rare kidney cancer. Her mother, Michelle, was tormented, as the parents of children stricken by cancer must always be, by the "Why my child?" question. She met Paul Brodeur in a San Diego bookstore owned by Brodeur's wife. It was the spring of 1990, and Brodeur was there promoting his new book, *Currents of Death*. The cause of little Mallory's tumor, he told Michelle, must surely be EMF. He introduced her to Michael Withey, a Seattle lawyer who had tried EMF cases. A year later Michelle and her husband Ted filed a lawsuit in San Diego Superior Court alleging that EMF from nearby San Diego Gas & Electric transmission lines was the cause of Mallory's cancer and had compelled them to sell their home at a loss.

EMF seemed to be a tort lawyer's dream. Withey had organized law firms across the United States into a group calling itself the Electromagnetic Radiation Case Evaluation Team that screened potential cases and maintained a computerized data bank. They saw the potential for a mass tort blitz. The mass tort industry draws its strength from sheer numbers. If they could persuade hundreds or thousands of frightened clients to file lawsuits all across the country, scientific evidence would become almost irrelevant. The prospect of simultaneously defending themselves from thousands of lawsuits would force power companies to reach a settlement. This a tactic had worked before: The asbestos onslaught a decade earlier had produced more than two hundred thousand lawsuits. No industry can deal with that.

Withey also organized a citizen-action group called the EMF Alliance that worked with the lawyers. It gave a grassroots cover to the search for people willing to file lawsuits against the power companies. The featured speaker at meetings of the EMF Alliance was Paul Brodeur, who could be counted on to cultivate paranoia over the unseen, unfelt menace of power lines. All that was needed now was a couple of big wins in court and the EMF blitz would take off on its own. Little Mallory was the poster child.

A front-page story in the *Wall Street Journal* quoted Withey as boasting that the case of four-year-old Mallory was "a slam dunk." He predicted it would trigger an avalanche of claims, and most legal experts agreed. The cover of the *Journal of the American Bar Association* showed an electric plug about to be inserted into an outlet. The title was "Plugging In: Why Electromagnetic Field Litigation Could Be the Next Asbestos." The "villains" were easy targets: wealthy power companies, government bureaucrats, and arrogant scientists. Moreover, the article observed, "public concern over EMF is rising irrespective of its validity." The best grounds for damage claims, therefore, would be loss of property value, in which case it would not be necessary to prove that there was a hazard, but only that there was a widespread perception of a hazard.

That's clearly what the lawyers for Martin and Joyce Covalt had in mind when they too filed suit against San Diego Gas & Electric in 1993. The Covalts did not claim that anyone in their family had been harmed by electromagnetic fields from nearby power lines; they contended that the fields had rendered their luxury home in San Clemente uninhabitable. It looked as though the expected avalanche of EMF litigation would start in San Diego. But a seemingly unrelated case before the highest court in the land was about to shift the odds.

SCIENTIFICALLY VALID PRINCIPLES

Daubert v. Merrell Dow Pharmaceuticals involved Bendectin, a morning-sickness drug that had been alleged to cause birth defects. Bendectin, the only medication approved by the Food and Drug Administration for the treatment of morning sickness, had been used by millions of women. More than thirty published studies involving more than 130,000 patients had found no measurable increase in birth defects. Nevertheless, lawyers for the parents of two children born with serious birth defects managed to find eight experts willing to testify, for a fee, that Bendectin might cause birth defects. There are, unfortunately, few scientific claims so far-fetched that no Ph.D. scientist can be found to vouch for them.

Daubert was archetypical of what has come to be known as junk science: science introduced in a court of law that would be scoffed at by most scientists. Junk science is quite distinct from the pathological science we dealt with in chapter 1. Pathological science results from scientists fooling themselves. Junk science is more sinister; it is deliberately designed to fool or befuddle nonscientists, particularly on juries. Typically, junk science consists of far-fetched or implausible scientific interpretations that are not supported by scientific evidence. It seeks to exploit the difficulty juries have in evaluating technical arguments.

In 1993 the Supreme Court ruled that such testimony is not credible and instructed federal judges to serve as "gatekeepers," aggressively screening out ill-founded or speculative scientific theories. Impressive credentials as an expert, the Court made clear, are not enough; evidence presented must be based on "scientifically valid principles."

The decision was a carefully crafted balancing act. On one side, it took into account the warnings of prominent scientists that peer review of published work does not ensure that it is correct. These scientists reminded the court that the history of science is a record of changing scientific consensus and argued that excluding as invalid evidence that is not "generally accepted" would sanction a scientific orthodoxy. On the other side, it was argued that allowing evidence to be introduced that is *not* generally accepted by the scientific community would result in a free-for-all in which juries would be confounded by absurd and irrational pseudoscientific assertions. By allowing evidence based on scientifically valid principles to be admitted even if it does not represent a scientific consensus, the Court found a middle course.

It was not at all clear that *Daubert* would have much effect in holding back the tide of junk science claims. It gave judges little guidance in how to decide whether or not evidence is based on "scientifically valid principles." Recognizing that judges are not scientists, the decision, written by Justice Harry Blackmun, invited district court judges to experiment with ways to fulfill their role as gatekeepers by seeking dispassionate scientific advice from outside the court. A few did.

In Oregon, federal district court judge Robert Jones responded to the *Daubert* decision by appointing a special panel of four independent scientists to evaluate expert testimony in some seventy cases involving silicone breast implants. He cited the opinion of Justice Blackmun in *Daubert* that district court judges must ensure that scientific testimony is "not only relevant but reliable." It was a recognition that neither judges nor juries are in a position to determine whether evidence is based on scientifically valid principles. The court, Judge Jones was saying, is more likely to arrive at the truth if it has the advice of scientific experts who have no stake in the outcome.

Silicone implants, which were introduced in 1962, were at one time widely used for cosmetic purposes but are used today almost exclusively for reconstruction after breast surgery. As many as a million American women have had implants. Thousands of them have sued the implant manufacturers, claiming that leaking implants caused illnesses ranging from cancer and such serious autoimmune diseases as lupus to vague symptoms such as headaches and fatigue. There is no question that it's possible for implants to rupture, allowing silicone to escape into the body, but silicone had been chosen for implants because it was believed to be biologically inert. For a fee, however, plenty of experts could be found to testify that leaking silicone *might* have caused the symptoms.

People who suffer a serious illness are often tormented by the "Why me?" question. It is natural for women who have had an implant to attribute subsequent ailments to the implant. It is simply the belief engine at work—*B* followed *A*—and the belief that *A* must therefore have caused *B* is strongly reinforced by the publicity surrounding the issue. It could be that *A* caused *B*, or it could be that there is no connection at all; such ailments also afflict many women who have never had an implant. The issue, then, is whether women who have had implants are more likely than others to suffer from these problems. It is a question that can only be answered by examining the statistics. In her book *Science on Trial*, Marcia Angell, the editor of the *New England Journal of Medicine*, who has carefully followed the court cases, concludes that jurors have difficulty "thinking in terms of probabilities or acknowledging the possibility of coincidence."

At least one implant manufacturer, Dow Corning, was driven into bankruptcy following court judgments as high as $25 million. It was not the high judgments that forced the bankruptcy, however, so much as the impossibility of defending against hundreds of lawsuits simultaneously. The real cost is human: women with implants who live in constant and unwarranted fear of complications, and women who would benefit from implants but refuse the procedure out of fear generated by the widespread publicity.

The testimony of the "expert" witnesses in the cases pending before the district court in Oregon consisted for the most part of far-fetched theories of how silicone gel might cause the sort of symptoms experienced by the women. The four disinterested scientists on the special panel advised Judge Jones that such theories, unsupported by peer-reviewed scientific data, simply were not credible. Following the panel's advice, the judge ruled that the plaintiff's lawyers could not introduce the expert testimony. There was simply no persuasive evidence in the testimony to indicate that women with leaking breast implants were at a greater risk of illness than other women.

Judge Sam Pointer Jr., who coordinates all breast implant cases in federal courts, took the idea of an independent panel a step further. The four scientists he appointed not only studied all the relevant published research but were allowed to question expert witnesses chosen by both sides. The report of Judge Pointer's panel, issued in December 1998, is expected to have a major impact not only on the thousands of breast implant lawsuits but on other junk science cases as well by encouraging other judges to rely on independent scientific panels.

Federal judges have always had the authority to seek independent advice from experts. In Joe Newman's suit against the Patent and Trademark Office, you will recall from chapter 5, Judge Thomas Penfield Jackson called in a special master to advise the court. And when the opinion of the special master seemed dubious, he turned to academic scientists for advice and eventually ordered Newman to turn the Energy Machine over to the National Bureau of Standards to be tested. He chose to act as a gatekeeper. What the *Daubert* decision did was to instruct federal judges that it is their *responsibility* to guard the gate.

THE AVALANCHE THAT NEVER HAPPENED

Although the Supreme Court decision in *Daubert* applied to federal courts, its influence is felt at the state level as well. Against this background, the California Supreme Court took up the Covalt lawsuit against San Diego Gas & Electric. It was the highest judicial body that had considered a case involving the alleged EMF health hazard. Specifically the court was asked to decide whether the Covalts were entitled to a jury trial or whether, as the power company insisted, their claim should be settled by the California Public Utilities Commission. This was a serious threat to the planned mass tort blitz. Tort lawyers thrive on jury trials; no tort lawyer wants to argue a junk science case before regulators.

Citing the U.S. Supreme Court decision in *Daubert v. Merrill Dow*, the California Supreme Court assumed the gatekeeper role and undertook its own thorough review of the science. But the California Supreme Court had no need to appoint an independent panel of experts to screen the evidence. The hundreds of published papers on EMF, including epidemiology, laboratory research, and theoretical analysis, had already been predigested by the American Physical Society, as we saw in the last chapter. The process of consensus building in the scientific community was well underway.

The California Supreme Court ruled against the Covalts, concluding that EMF cases "have no place in the courtroom." The seventy-two-page decision was written by Justice Stanley Mosk, a liberal judge not noted for his sympathy to industry. It could have served as a textbook on the interaction of EMF with the human body. The Covalts had contended that "there have been many positive studies of EMF-cancer reported in the scientific literature," yet they cited only one such study: the 1995 study of U.S. electrical workers that found a possible risk of just one rare form of brain cancer. Justice Mosk observed that, by contrast, there were "noteworthy expressions of consensus" among scientists saying there was no EMF-cancer link. It was a reference not only to the reviews by the Oak Ridge Associated Universities and the American Physical Society but also to an amicus curiae (friend of the court) brief filed by a group of prominent scientists, including physicists, biologists, epidemiologists, and medical researchers. I was one of the sixteen who signed the brief. Six of our group were Nobel laureates. None of us had any conceivable connection to the power industry. The *Covalt* decision effectively ended EMF litigation in California and dampened the enthusiasm for such cases nationwide. It was a factor in the failure of similar suits in Florida and Texas.

Eight months after the *Covalt* decision, the National Academy of Sciences report on the health effects of residential electromagnetic fields confirmed what the scientists had been saying all along. A few months later the National Cancer Institute study extinguished any serious doubts that remained. The danger of an avalanche had passed. In spite of the predictions of the tort lawyers, not a single suit based on health effects from EMF ever succeeded. The parents of Mallory, the little girl with kidney cancer, lost their suit, but Mallory won her battle against cancer and is fully recovered.

THE COURT REVISITS JUNK SCIENCE

It had been expected that the Supreme Court would have to revisit *Daubert* at some point to give judges more guidance in fulfilling their role as gatekeepers. Having invited experimentation, the Supreme Court would have to take up a lower court decision that had been based on *Daubert* and decide what the experiment had learned.

The case they chose was *General Electric v. Robert Joiner.* Robert Joiner began working as an electrician in the Water and Light Department of Thomasville, Georgia, in 1973. His work involved frequent contact with transformer fluid, which sometimes splashed into his eyes and mouth. Ten years later, the city discovered that some of the transformers were contaminated with a residue of polychlorinated biphenyls (PCBs). PCBs were once widely used as a fire-resistant dielectric fluid in transformers, but their production and use are now banned in most industrialized nations because of their toxicity and suspected carcinogenicity. Joiner was a heavy smoker, his parents were both heavy smokers, and there was lung cancer in his family, but when he developed lung cancer in 1991, he blamed the PCBs, and sued General Electric, the company that manufactured the transformers.

Joiner's lawyers apparently had no trouble finding "experts" who were willing to testify that the minute PCB contamination had contributed to his cancer. Their testimony was based on studies of infant mice that had massive doses of PCBs injected directly into their stomachs. The mice developed cancer, all right, but of a form unrelated to lung cancer. It seemed to be a classic example of junk science, and a district court judge, citing *Daubert*, dismissed the expert testimony. However, an appeals court reinstated the testimony on the grounds that *Daubert* only required that the evidence be based on scientifically valid principles. The evidence presented by Joiner's lawyers was scientific enough — it just didn't have much to do with Joiner's cancer.

On December 15, 1997, the Supreme Court, in a unanimous decision, reversed the appeals verdict in *General Electric Co. v. Robert Joiner,* ruling that the district court judge had acted properly in requiring that conclusions drawn from the evidence must make scientific sense. The effect was to greatly strengthen *Daubert*: not only must evidence be obtained by scientifically valid procedures, it must also be scientifically interpreted.

In the long run, a concurring opinion by Justice Breyer may have an even greater impact. Again noting the obvious, "judges are not scientists," Breyer encouraged trial judges to appoint independent experts to serve on behalf of the court. He noted that courts can turn to scientific organizations such as the National Academy of Sciences and the American Association for the Advancement of Science to identify neutral experts, not to resolve the scientific issues, but to keep information that is unsound, unhelpful, and unreliable away from the jury.

Many district court judges, however, continue to feel that *Daubert* is a last resort, to be used only when the adversary system fails. The problem is deciding when the system is failing. In the name of "impartiality," juries today are often deliberately comprised of the least informed among us. There is a mythology among judges that a group of twelve incurious and ill-read citizens will have some special ability to discern the truth. It is the myth of the human lie detector: somehow jurors will sense from shifty eyes or nervous speech who is telling the truth. A scientist asks for the evidence of this remarkable power. We really have no way of knowing what goes on in the jury room, and unless someone devises a metric by which to measure the performance of juries, there is reason to be concerned about their ability to sensibly resolve technical issues.

The *Daubert* and *Joiner* decisions represent a major advance in dealing with junk science, and as sitting judges retire, their replacements will most likely invoke them more readily. Additional Supreme Court decisions will probably be needed to further refine *Daubert* and *Joiner*. Voodoo science has a way of winding up in the courts, but at least shark repellent is now available.

Junk science raises a more troubling concern for the scientific community. The pathological science we encountered in earlier chapters involved scientists fooling themselves. Pseudoscience often involves the tendency to fill in scientific uncertainty with views based on political or religious convictions. In both cases, scientists may be mistaken or even foolish, but it can be argued that, at least in the beginning, they believe their claims to be true.

Science is grounded in the assumption that there is no deliberate attempt to deceive. But in junk science, we contend with scientists, many of whom have impressive credentials, who craft arguments deliberately intended to deceive or confuse. And yet it does not generally rise quite to the level of fraud. Rather, it often consists of convoluted theories of what *could* be so, with little or no supporting evidence. Since such theories are not normally published in the open scientific literature or presented at scientific conferences, junk science can exist entirely outside the realm of scientific discourse, immune from the self-correcting mechanisms of genuine science.

Junk science is an example of voodoo science that survives by avoiding the full scrutiny of the scientific community. In the next chapter we will consider a much larger body of scientific research that is kept hidden: every government believes that science related to national security must be withheld from full scientific scrutiny and debate. It should be no surprise that in such an environment of official secrecy voodoo science flourishes.

NINE
ONLY MUSHROOMS GROW IN THE DARK
In Which Voodoo Science Is Protected by Official Secrecy

THE ROSWELL INCIDENT

IN 1994 SECRETARY of the Air Force Sheila Widnall agreed to issue an unprecedented blanket order relieving anyone with information about an alleged 1947 UFO incident near Roswell, New Mexico, from any obligation to keep the information secret. A no-nonsense physicist and aeronautical engineer on leave from MIT, Secretary Widnall thought the air force had more important business than chasing down UFO stories, but Representative Steven Schiff of New Mexico was insisting on an all-out search for records and witnesses. Schiff wanted to reassure the public that there was no government cover-up. I did not expect anyone to come forward with new information, but I recalled with some chagrin my own "Roswell incident."

As a young air force lieutenant in the summer of 1954, I had been sent on temporary assignment to Walker Air Force

Base in Roswell to oversee the installation of a new radar system. I was returning to the base after a weekend visit with my family in South Texas. It was after midnight, and I was driving on a totally deserted stretch of highway in one of the most desolate regions of West Texas. It was a moonless night but very clear, and I could make out a range of ragged hills off to my left, silhouetted against the background of stars. Suddenly the entire countryside was illuminated by a dazzling blue-green light streaking across the sky just above the horizon. It flashed on and off as it passed behind the hills—and vanished without a sound. It was all over in perhaps two seconds. It came at a time when reports of UFO sightings were in the news almost daily. Most of the reported sightings were not nearly so spectacular as the event I had just witnessed, but I was pretty sure I knew what this one was.

The pale blue-green color is characteristic of the light emitted by frozen free-hydroxide radicals as they warm up. A free radical is a fragment of a molecule; the hydroxide, or OH, radical is a water molecule that is missing one of its hydrogen atoms. Free radicals are highly reactive, anxious to reconnect with their missing parts, and do not ordinarily stick around very long. However, if molecules are broken up by radiation at very low temperatures, the fragments can be frozen in place and unable to recombine. As soon as the severed parts of the molecule are warmed up, they react with other fragments to form stable molecules. The law of conservation of energy, once again, tells us what to expect: the energy it took to break the chemical bonds in the first place will be liberated when the fragments recombine. The liberated energy appears as blue-green fluorescence. An ice meteoroid, traveling through the cold depths of space for eons, will gradually accumulate more and more free radicals as a result of cosmic ray bombardment. I must have been fortunate enough to see an ice meteorite as it plunged into the upper atmosphere. The meteorite must have evaporated without a trace before reaching the ground.

As I continued down the highway and crossed into New Mexico, I was feeling rather smug. The UFO hysteria that was sweeping the country, I told myself, was for people who don't understand science. It was then that I saw the flying saucer. It was again off to my left between the highway and the distant hills, racing along just above the range land. It appeared to be a shiny metallic disk viewed on edge—thicker in the center—and it was traveling at almost the same speed I was. Was it following me? I stepped hard on the gas pedal of the Oldsmobile—and the saucer accelerated. I slammed on the brakes—and it stopped. Then I could see that it was only my headlights, reflecting off a single phone line strung parallel to the highway. Suddenly, it no longer looked like a flying saucer at all.

It was humbling. My cerebral cortex may have sneered at stories of flying saucers, but the part of my brain in which those stories were stored had been activated by the powerful impression of the ice meteorite. The belief engine did the rest. I was primed to see a flying saucer—and my brain filled in the details. Whenever I become impatient with UFO believers, as I often do, I try to remember that night in New Mexico when, for a few seconds, I believed in flying saucers.

ABDUCTED

Who has not seen in the dusk an animal that turns into a bush as you grow closer? But something more than the mind playing tricks with patterns of light is needed to explain why hundreds, by some accounts thousands, of people claim to have been abducted by aliens, taken aboard a spaceship, and subjected to some sort of physical examination, usually focusing on their erogenous zones. The examination is frequently followed by the aliens inserting a tiny implant into the abductee's body. Often the memory of these abductions has a dreamlike quality, and the subjects are able to recall the details only under hypnosis.

For these people, space aliens are a serious reality, but it's not clear how much science can help. Indeed, scientists themselves are not immune to such beliefs. In 1992 a five-day conference was held at MIT to assess the similarities among various alien abduction accounts. The conference was organized by John Mack, a professor of psychiatry at Harvard, and David Pritchard, a prizewinning MIT physicist. Mack had been treating patients who believed they had been abducted. His treatment consisted of assuring them that they were not suffering from hallucinations—they really had been abducted.

Pritchard, an experimentalist, was more concerned with examining any physical evidence, particularly the tiny implants many abductees reported. The best candidate seemed to be an implant that abductee Richard Price said had been inserted midshaft in his penis. The implant was clearly visible, amber in color, the size of a grain of rice. Under a microscope, what appeared to be wires could be seen protruding from the device. What wonders of alien technology might be revealed by this tiny device? In an atmosphere of high expectations, the "implant" was removed and examined by sophisticated analytical techniques. It was not from Andromeda. It was of distinctly terrestrial origin: human tissue that had accreted fibers of cotton from Price's underwear.

It is hardly surprising that there are similarities in the accounts of the abductees; they've all been exposed to the same images and stories in the popular media. In my local bookstore, there are three times as many books about UFOs as there are about all of science. Aliens stare at us from the covers of magazines and appear in television commercials. They are the subject of hundreds of movies and television series—even "documentaries," if you want to call them that.

As time goes by, there is a growing uniformity in the descriptions of aliens. A six-year-old child can now sketch what an alien looks like. We are, in fact, witnessing a sort of alien evolution. The mutations are created by moviemakers and science fiction novelists. The selection mechanism is public approval. The aliens subtly evolve to satisfy public expectations, resulting in a sort of composite alien. In effect, the public is voting on what aliens should look like. The same holds true for UFOs.

If you attempt to trace the process back to a common ancestor, the trail seems inevitably to lead back to the strange events near Roswell, New Mexico, in the summer of 1947.

PROJECT MOGUL

On June 14, 1947, William Brazel, the foreman of the Foster Ranch, seventy-five miles northwest of Roswell, spotted a large area of wreckage about seven miles from the ranch house. The debris consisted of neoprene strips, tape, metal foil, cardboard, and sticks. He didn't bother to examine it very closely at the time, but a few weeks later he heard about the first reports of flying saucers and wondered if what he had seen might be related. He went back with his wife and gathered up some of the pieces. The next day, he drove to the little town of Corona to sell wool, and while he was there he "whispered kinda confidential like" to the Lincoln County sheriff, George Wilcox, that he might have found pieces of one of these "flying discs" people were talking about. The sheriff reported it to the army air base in Roswell. The army sent an intelligence officer, Major Jesse Marcel, to check it out. Marcel thought the debris looked something like pieces of a weather balloon or a radar reflector. All of it together fit easily into the trunk of his car.

It might have ended there, but the Public Information Office at Roswell Army Air Field issued a garbled account to the press the next day saying the army had "gained possession of a flying disc through the cooperation of a local rancher and the sheriff's office." The army quickly issued a correction describing the debris as a standard radar target. It was too late. The Roswell incident had been launched. With the passage of years, the retraction of that original press release would come to look more and more like a cover-up.

When I was sent to Roswell a few years later to install the new radar, Roswell Army Air Field had been renamed Walker Air Force Base. It was home to a force of B-36 long-range bombers. The Soviet Union had the bomb, and a rapid buildup of our strategic forces was underway. The bachelor officers' quarters on base were filled up when I arrived, so I took a room in town, in a large boardinghouse on a pleasant street lined with cottonwoods.

The other residents of the boardinghouse, all Roswell natives, were much older. It was almost like a family, but they went out of their way to make me feel at home, chiding me good-naturedly about all the "secret stuff" going on at Walker. One balmy July evening on the front porch, the conversation turned to flying saucer stories. They knew about the wreckage found on the Foster Ranch in 1947—it had all been in the *Roswell Daily Record*. Not a single person believed the government explanations about weather balloons or radar targets; everyone seemed to agree the debris must have been from a secret government project, or maybe some sort of experimental Russian aircraft. I do not recall anyone suggesting it was from outer space.

It was not until 1978, thirty years after William Brazel spotted wreckage on his ranch, that alien bodies first showed up in accounts of the "crash." The story of Major Marcel loading sticks, cardboard, and metal foil into the trunk of his car had grown over the years into a major military operation to recover an entire alien spaceship that was secretly transported to Wright-Patterson Air Force Base in Ohio. Even as the number of people who might recall events thirty years earlier dwindled, incredible new details began to be added by second- and third-hand sources: there was not one crash but two or three; the aliens were small with large heads and suction cups on their fingers; one alien survived for a time but was kept hidden by the government; and on and on.

Like a giant vacuum, the story had sucked in accounts of un-related plane crashes and high-altitude parachute experiments using anthropomorphic dummies, even though in some cases these events took place years later and miles away. Various UFO "inves-tigators" managed to stitch together fragments of these accounts to create the myth of an encounter with extraterrestrials—an en-counter covered up by the government. The truth, according to believers, was too frightening to share with the public.

If the pieces didn't fit, they were trimmed until they did. If they couldn't be made to fit, they were left out. To fill the huge gaps that remained, the faithful speculated. In time, the distinction be-tween fact and speculation faded. A string of profitable books was generated, and then a string of skeptical responses by aerospace writer Philip Klass. It is an axiom in the publishing business, how-ever, that pseudoscience will always sell more books than the real science that debunks it.

Roswell was a gold mine. The unverified accounts were shame-lessly exploited for their entertainment value on television pro-grams that represented themselves as documentaries, such as NBC's *Unsolved Mysteries* with host Robert Stack, and even more serious news programs, such as CBS's *48 Hours* with Dan Rather, not to mention talk shows, including CNN's *Larry King Live*.

The bottom was reached by Fox TV, which in 1995 showed grainy black-and-white film of what was purported to be a government autopsy of one of the aliens. The film was immediately denounced by experts as an obvious hoax, but it scored high ratings with the viewing public. The experts, people shrugged, were probably paid off or threatened by the government. Fox continued showing the film over and over.

When the ratings for *Alien Autopsy* finally began to slip after three years, Fox announced that it had hired its own experts to examine the film. Using high-tech "NASA-type video enhancements," they revealed the shocking truth: the film was a fake. Was Fox chagrined at having been duped? Not at all. Fox boasted of having exposed "one of the biggest hoaxes of all time." A highly promoted special was aired that described how the autopsy film had been faked. Fox had managed to make a profit from the Roswell incident coming and going.

Meanwhile, however, to the astonishment of believers and skeptics alike, the search of air force records for information about the Roswell incident uncovered a still-secret government program from the 1940s called Project Mogul. There really was a cover-up — but not of an alien spaceship.

In the summer of 1947, the Soviet Union had not yet detonated its first atomic bomb, but it was clearly only a matter of time. It was imperative that the United States know about it when it happened. A variety of approaches to detect that first Soviet nuclear test was being explored. Project Mogul was an attempt to use low-frequency acoustic microphones placed at high altitude to actually "hear" the explosion. The interface between the troposphere and the stratosphere creates an acoustic "duct" that can propagate sound waves globally. Acoustic sensors to pick up the explosion, radar tracking reflectors, and other equipment was sent aloft on trains of weather balloons as long as six hundred feet.

The balloon trains were launched from Alamagordo, New Mexico, about a hundred miles west of Roswell. One of the surviving scientists from Project Mogul, Charles B. Moore, a retired physics professor, recalls that Flight #4, launched on June 4, 1947, was tracked to within seventeen miles of the spot where William Brazel spotted the wreckage ten days later. At that point, contact was lost.

The debris found on the Foster Ranch closely matched the materials used in the balloon trains. The air force now concludes that it was, beyond any reasonable doubt, the crash of Flight #4 that set off the bizarre set of events known as the Roswell incident. Had Project Mogul not been highly secret, unknown even to the military authorities in Roswell, the entire episode might have ended in July 1947.

From today's perspective, it is difficult to understand why Project Mogul was secret at all. It was abandoned even before the Soviets tested their first atomic bomb, pushed aside by more promising detection technologies. There was nothing in Project Mogul that could have provided the Soviets with anything but amusement, and yet it was a tightly kept secret for nearly half a century; even its code name was secret. It would still be secret if it had not been for the investigation initiated by Representative Schiff. Secrecy, it seems, is simply a part of the military culture, and it has produced a mountain of secret materials.

No one really knows the size of the classified mountain, but in spite of periodic efforts at reform, there are more classified documents today than there were at the height of the cold war. The direct cost of maintaining them is estimated by the government to be about $2.6 billion per year, but the true cost in terms of the erosion of public trust is immeasurable. In a desperate attempt to bring the system under control, President Clinton issued an executive order in 1995 that will automatically declassify documents more than twenty-five years old—estimated at well over a billion pages—beginning in the year 2000.

If there is any mystery still surrounding the Roswell incident, it is why uncovering Project Mogul in 1994 failed to put an end to the UFO myth. There appear to be several reasons, all related to the fact that the truth came out almost half a century too late. Rather than weakening the UFO myth, Project Mogul was pounced on by believers as proof that everything the government had said before was a lie, and there was no reason to believe this was not just another lie. Government denials are by now greeted with derision.

But if it was Project Mogul that started the UFO myth, it was another secret government program that kept it going. It was common during the cold war to create cover stories to protect secret operations, including flights of the U-2 spy plane over the Soviet Union. Initially, the U-2s were unpainted; that is, their skin was shiny, metallic aluminum, which strongly reflected sunlight. Particularly in the morning and evening hours, when the surface below was dark, the U-2s would pick up the Sun's rays, becoming very visible. The CIA estimates that over half of all UFO reports from the late 1950s through the 1960s were secret reconnaissance flights by U-2 spy planes. To allay public concern while maintaining the secrecy of the U-2 missions, the air force concocted far-fetched explanations in terms of natural phenomena. Keeping secrets, we learn early in life, leads directly to telling lies.

The U.S. Air Force collected every scrap of information dealing with the Roswell incident into a massive report in hopes of bringing the story to an end. The enormous task of locating and sifting through old files and tracking down surviving witnesses had actually started even before Representative Schiff's call for full disclosure. Responding to Freedom of Information Act requests from self-appointed UFO investigators had become a heavy burden on the air force headquarters staff at the Pentagon, and they were eager to get ahead of the Roswell incident. Release of *The Roswell Report: Case Closed* drew the largest attendance on record for a Pentagon press conference.

Although the people involved insist it was not planned that way, the air force report was completed just in time for the fiftieth anniversary of William Brazel's discovery of the Project Mogul wreckage. Thousands of UFO enthusiasts descended on Roswell, now a popular tourist destination, in July 1997 for a golden anniversary celebration. They bought alien dolls and commemorative T-shirts and snatched up every book they could find on UFOs and aliens. The only book that sold poorly was the massive air force report. Who, after all, could take the government seriously? Fox TV continued to show its alien autopsy film to appreciative audiences. Recent polls indicate that the number of people who believe there is a UFO presence that is being covered up by the government is still growing.

Nevertheless, it is easy to read too much significance into reports of widespread public belief in UFOs and alien visits to Earth. Carl

Sagan saw in the space-alien myth the modern equivalent of the demons that haunted medieval society, and for a susceptible few they are a frightening reality. But for most people these do not seem to be deeply held beliefs. UFOs and aliens are a way to add a touch of excitement and mystery to uneventful lives. They're also a way for people to thumb their noses at the government.

The real cost of the Roswell incident must be measured in terms of the loss of public trust. In the name of national security, every government in this troubled world feels compelled to grant itself the authority to hold official secrets. Those in power quickly learn to love secrecy. It enables the government to control what the public hears: bad news is squelched, good news is leaked. In the long run, however, episodes like Roswell leave the government almost powerless to reassure its citizens in the face of far-fetched conspiracy theories and pseudoscientific hogwash.

The release of *The Roswell Report: Case Closed* on June 24, 1997, came just three months after the bodies of thirty-nine members of a UFO cult called Heaven's Gate were found in San Diego. They had committed mass suicide in the belief that a giant UFO following the Hale-Bopp comet would pick them up and carry them to the "next level." Behind the curtain of official secrecy, however, far more dangerous deceptions have gone undetected. Consider the case of Star Wars and the X-ray laser.

STAR WARS

In November 1988 the conservative elite of the United States gathered in Washington to honor physicist Edward Teller, the legendary "father of the H-bomb," as "a patriot who has combined profound moral judgment with political wisdom." An award for his contributions to the Strategic Defense Initiative (SDI) was presented by Sanford McDonnell, CEO of McDonnell-Douglas, a major SDI contractor. President Ronald Reagan appeared on closed-circuit television to praise Teller as "a tireless advocate" of SDI.

Among those invited to pay tribute to Teller was Andrei Sakharov, the dissident Soviet physicist who had headed the Soviet development of the hydrogen bomb. Having already been awarded the Nobel Peace Prize in 1975, he was in the United States to receive the Einstein Peace Prize. Friends urged him to decline the invitation to speak at Teller's celebration, but true to his principles, Sakharov felt compelled to voice his concerns. Stooped and frail from the repeated hunger strikes that had been his only means of resistance during the years of exile in Gorky, Sakharov was clearly out of place in an ill-fitting beige suit among jewel-bedecked women and tuxedo-clad men. With his son-in-law serving as interpreter, he spoke respectfully of Teller but warned that Teller's support for SDI was a "grave error." Deployment of a space-based antimissile defense system, he argued, could destabilize the nuclear standoff—if it could be made to work at all, he added prophetically, before bankrupting the superpowers.

Sakharov left immediately following his remarks. As the door closed behind him, William F. Buckley Jr., who was serving as master of ceremonies, summed it up for the audience: "A nice man," he sniffed, "but clearly out of touch with reality." Then Teller took the podium. He began his speech by reminding the audience that the Kremlin had lifted Sakharov's security clearance twenty years earlier, leaving him hopelessly uninformed about modern developments in weapons. But he, Edward Teller, had continued to work in lasers and nuclear weapons. It was standard Teller. Although he publicly deplores secrecy, Teller repeatedly takes refuge behind it. "If only you knew what I know," he will say regretfully in his Hungarian accent. "I wish I could tell you."

Teller is endowed with a fertile imagination and a sense of the dramatic, but his charisma defies analysis. For half a century, with the threat of nuclear war shaping every world event, Teller and the H-bomb became almost synonymous. But for all his celebrity and genius, behind Teller stretches an almost unblemished record of technical failure, going back to his original concept for a thermonuclear bomb. Teller's version was eventually abandoned as impractical; the first H-bomb was based on the ideas of Stanislaw Ulam. But it was Teller, a master of public relations, who was publicly hailed as the "father of the H-bomb." The public was completely unaware of Ulam's contribution; only Teller's name was associated with the bomb.

In his book *Teller's War*, Bill Broad of the *New York Times* points out that the young Teller produced a continuous stream of unconventional ideas on an incredibly wide range of subjects, but few of his ideas were practical. He was productive only when he was teamed up with great physicists such as Hans Bethe and Freeman Dyson, who forced him to confront reality. But after his testimony in the Oppenheimer hearings in 1954, many leading physicists refused even to speak to him.

J. Robert Oppenheimer, the physicist who directed the development of the atomic, or fission, bomb, was revered by most of the scientists involved in the Manhattan project. After the war, however, Oppenheimer had serious reservations about the wisdom of developing a hydrogen, or fusion, bomb. Teller was the most conspicuous scientific proponent of a "superbomb." It was the height of the McCarthy era in Washington, and advocates of the H-bomb sought to portray Oppenheimer's opposition as evidence that he was a security risk, a charge most scientists regarded as preposterous. Teller stopped short of labeling Oppenheimer a Communist agent, but he testified in a hearing before the Atomic Energy Commission (AEC) that it would be "wiser" to deny Oppenheimer a security clearance. The AEC stripped Oppenheimer of his clearance. In 1954, under the provisions of the Atomic Secrets Act, that meant Oppenheimer was effectively barred from practicing his profession as a nuclear physicist. It also meant he could no longer oppose Edward Teller. Many physicists never forgave Teller for what they regarded as his betrayal of a talented and honorable colleague.

Isolated at the Lawrence Livermore Laboratory, which had been created for him by supporters in Congress, Teller surrounded himself with sycophants who rarely questioned his judgment. His isolation became almost total after he began collaborating with a brash young physicist named Lowell Wood. Far different from the cool-headed scientists that kept Teller on track in the early years, Wood overflows with outrageous ideas of his own. Teller gave Wood access to power; Wood provided renewal of the wellspring of ideas that had begun to dry up in Teller.

It was a disastrous collaboration. They share the same flaw: both suffer from brilliance untempered by judgment. Like Fleischmann and Pons, they reinforced each other's worst instincts, and in so doing fell victim to wishful thinking. The sad product of their collaboration was Star Wars and the X-ray laser. Like Pons and Fleischmann, they would seek to conceal their blunder. For Teller and Wood it was easy: the screen was provided by official government secrecy.

THE MYTHICAL X-RAY LASER

The Strategic Defense Initiative burst upon the science community without warning on March 23, 1984. President Reagan, in a speech to the American people, made a direct appeal to scientists: "I call upon the scientific community, those who gave us nuclear weapons, to turn their great talents now to the cause of mankind and world peace, to give us the means of rendering these weapons impotent and obsolete." Surely no president would launch the nation on a technological undertaking with such awesome implications as SDI, or Star Wars as it quickly came to be known, unless he had already been assured the prospects for success were good. If it failed, it would represent an enormous diversion of resources from the unmet needs of the nation. Even if it succeeded technologically, it could block a negotiated reduction in nuclear weapons or even destabilize the nuclear standoff. By promising to render one nation invulnerable to counterattack, an impenetrable defense would threaten the rest of the world with nuclear blackmail. The Soviet Union might have chosen to take its chances with a first strike of its own before such a system could be put in place.

To advise the president on just such questions, there is a White House Science Office, whose director is the science advisor to the president. The science advisor in 1984 was George Keyworth, a relatively young physicist of no very notable accomplishment from the Los Alamos nuclear weapons laboratory. Almost totally unknown to the scientific community, Keyworth was a far cry from the distinguished leaders of American science who had held the post in previous administrations. His name had been suggested to the president by Edward Teller, who knew Reagan from his days as governor of California. Keyworth was widely regarded as Teller's man in the White House. What did Keyworth think of SDI? No one thought to ask. He did not even learn of the president's speech until a couple of days before it was delivered. The White House astrologer may have had more influence on national policy.

It is Teller himself who is generally credited with being the key influence behind Ronald Reagan's Star Wars vision. Teller had met with the president a few months earlier to urge support for research on an X-ray laser driven by a nuclear explosion. Teller referred to the X-ray laser as a "third-generation" nuclear weapon. The atomic bomb was the first generation, and the hydrogen bomb the second; they simply produced explosions, radiating energy in all directions. The third generation would focus energy from a nuclear explosion into incredibly intense beams that could destroy targets at great distances. It would be, Teller argued, the perfect weapon to defend against enemy missiles. Reagan was easy to persuade. His "kitchen cabinet" of unofficial advisors had been urging creation of a missile defense system, which he saw as a simple moral issue. "Would it not be better to save lives than to avenge them?" he had asked in his Star Wars address.

But were the prospects for an X-ray laser as strong as Teller claimed? Anyone who knows Teller's record recognizes that he is invariably optimistic about even the most improbable technological schemes. He is almost the archetype of the scientist who loves technology too much. The driving force behind the X-ray laser was Teller's protégé, Lowell Wood, who like Teller is inclined to brush away daunting technical obstacles with a wave of his hand. In fact, the X-ray laser concept had never been tested. It was based on the theoretical work of a strange young scientist named Peter Hagelstein who had joined Wood's group. Wood, whose own theoretical skills were limited, was persuaded that Hagelstein was a genius, and Hagelstein soon began believing it too.

The idea behind the X-ray laser was simple enough. Atoms in an excited state emit light when they decay back to their ground state. The decay is normally random, but in a laser, radiation of the proper frequency stimulates the excited atoms to decay in unison. The result is a very intense and directional beam of photons. Depending on the energy of the excited state, the photon emitted in the decay can be in the infrared or the visible part of the spectrum. In principle, it can also be in the X-ray region, but that is extraordinarily difficult to achieve, since it is the deep-lying, or X-ray, levels of the atoms that must be stimulated. The plan was to use aluminum as the lasing material. Because aluminum has only a single X-ray level, the emitted photons would be concentrated at a single energy. The aluminum would be in the form of a rod that would generate a pencil-thin beam of X rays. Hagelstein's calculations indicated it would work if there was a sufficiently intense source of stimulating radiation to "pump" the laser—such as the radiation from a nuclear explosion.

So this was the "third-generation" nuclear weapon, an X-ray laser pumped by the radiation of a nuclear explosion. The radiation would reach the aluminum rod just ahead of the shock wave. The laser would fire its deadly beams of X rays and be vaporized an instant later. Well, that was the plan. It was given the code name Excalibur. To have named the X-ray laser after a mythical weapon would turn out to be singularly appropriate.

Stories were leaked to the press about the fearsome X-ray laser, including artists' renderings in full color. It became the central component, at least in the media hype, of the Strategic Defense Initiative. When questions were raised about the usefulness of a weapon that could only be fired once against an attack that might include thousands of missiles, *Super* Excalibur suddenly appeared on the scene as if from nowhere. Super Excalibur, according to the steady leak of top secret information, would bristle with aluminum rods capable of generating a hundred or a thousand separately targeted laser beams simultaneously. The leaked information gave little hint of the state of development, but members of the press tended to assume that to be at the center of an undertaking of the vast scope of SDI, it must have solid evidence behind it. Even Congress believed we were close to developing such a weapon—and so did the Soviets.

Two years and $8 billion later, the United States appeared to be on the verge of concluding a treaty with the Soviet Union calling for the elimination of offensive nuclear weapons within ten years—on the condition that SDI be confined to the laboratory. It seemed to be a triumph of the Reagan policy; SDI had brought the Soviets to the peace table. But that's not what Teller had in mind. On the eve of the Rekjavik summit in October 1986, Teller sent an urgent message to Paul Nitze, the chief U.S. arms negotiator, telling him that Super Excalibur had been successfully tested and was "ready for engineering development." A single X-ray laser, "the size of an office desk," Teller boasted, could potentially shoot down the entire Soviet land-based missile force. His message was clear: don't make any deals; we have the upper hand. There is no way to know what impact, if any, Teller's message had, but the Rekjavik summit, which had seemed to hold such promise for world peace, ended in total failure.

And what of the fearsome Super Excalibur? There never was an X-ray laser. There had been a single test of the concept—it ended in confusion. To appreciate the level of confusion, you need to know how such tests are conducted. A nuclear bomb was detonated at the bottom of a deep vertical shaft drilled in the Nevada desert. Just above the bomb was the laser unit, with instruments to detect any X rays the laser might emit. There was telemetry to relay information at the speed of light from the detectors to receiving stations at the surface. Finally the shaft was filled in. Any information had to be collected in the instant between the arrival at the laser of X rays from the blast, which travel at the speed of light, and the shock wave from the blast, which travels at the speed of sound. When the shock wave arrives, the laser, the detectors, the telemetry are all obliterated. You get just one shot. There is no way to go back and check to see if things were working properly.

Initial results seemed to indicate that some lasing had taken place. Wood's group at Livermore held a celebration, and word was leaked far and wide that the test had been a success. More careful analysis, however, revealed that what had been interpreted as laser action may, in fact, have been oxygen fluorescence from a faulty detector. Super Excalibur had not produced a beam of X rays powerful enough to knock down a butterfly. The truth was buried, as failures usually are, by top secret classification, but there would never be another test of the dreaded Super Excalibur.

Although neither Congress nor the public was aware of any of this, many scientists had from the beginning questioned whether so-called directed-energy weapons (DEW) were feasible. In 1984, a year after President Reagan's Star Wars speech, I was one of a small group from the American Physical Society who paid a visit to George Keyworth, the president's science advisor. We proposed to convene a group of acknowledged experts to examine the feasibility of DEW.

Keyworth embraced the idea and arranged for us to meet with General James Abrahamson, who had been placed in charge of the SDI program. The study could not be conducted without full access to classified information, and only Abrahamson could authorize that. Abrahamson agreed to cooperate fully and even arranged for top secret briefings from various laboratories that were working on the project. It took about a year to assemble the study panel; it was headed jointly by Professor Nicolass Bloembergen of Harvard, who won the Nobel Prize in 1981 for his research with lasers, and Dr. Kumar Patel, the director of research at Bell Laboratories and the inventor of the CO_2 laser. The distinction of these two scientists, as well as the other sixteen members, ensured that the findings of the panel would be taken seriously.

The study took eighteen months and was held up by Pentagon censors for another seven, but finally, on April 23, 1987, three years after President Reagan's Star Wars speech, an unclassified version of "The Science and Technology of Directed Energy Weapons" was released to the public. The panel concluded that at least ten years would be needed just to determine if the concept was feasible—and it did not look promising. This study marked the turning point for Star Wars. Funding for the program gradually wound down over the next few years.

All work on the X-ray laser concept ended in March 1988. What became of Wood and Teller? They were back in Washington. Without comment or apology, they had simply replaced Super Excalibur with Wood's latest ultimate weapon: small, cheap, self-contained, space-based interceptors that would hurl themselves into the path of an enemy missile. When they were derisively referred to as "smart rocks," Wood adopted the name "Brilliant Pebbles." At least one skeptical congressman insisted on referring to them as "loose marbles."

Apparently untroubled by this bait-and-switch tactic, national leaders who had fallen for the X-ray laser story now embraced Brilliant Pebbles with the same fervor. It was an amazing demonstration of Teller's ability to remain standing amid the rubble of his ideas, but it was not enough to save the Strategic Defense Initiative. On May 14, 1993, SDI was officially terminated. Although by then $30 billion dollars had been spent on SDI, there was almost nothing to show for it; the money had simply been swallowed up by defense contractors.

Star Wars, of course, came during a period of intense cold war confrontation. In the interests of security, people in every society grant their government a license to keep secrets; in times of perceived national danger, the license is broadened. It is a necessary but dangerous bargain. Behind the curtain of official secrecy, waste, corruption, and foolishness can be concealed, and information can be selectively leaked for political advantage. In the case of the Strategic Defense Initiative, erroneous scientific information and flawed technical concepts, protected from normal scientific scrutiny, were used to promote costly defense programs that in the end only left the United States more vulnerable. Secrecy had, as it so often does, provided a haven for voodoo science.

Proponents of SDI today claim the program was really a clever deception that hastened the collapse of the Soviet Union. That it contributed to the collapse, as Sakharov suggested it might, is no doubt true. There is, however, no evidence that it was part of some strategic plan. If it was, it was a singularly dangerous plan that could have ended with a catastrophic response by threatened Soviet leaders. The cost that cannot be measured is the loss of the people's trust in their government.

There are, of course, many superb scientists engaged in secret research. In this imperfect world, it's a necessity. Unfortunately, even the best scientists are hampered if they must forego the critical feedback of the full scientific community. By contrast, as we saw particularly in the case of cold fusion, scientists engaged in questionable research consciously or unconsciously seek to isolate themselves from critics. Official secrecy makes it easy; it offers a refuge for incompetence. Not only does secrecy contribute to scientific blunders, it allows those blunders to remain hidden. Nowhere was that more true than in France during the energy crisis.

THE SNIFFER PLANE

In 1976 a Belgian count persuaded the government of France to conduct trials of a secret device that purportedly used the echo from a newly discovered particle to map mineral deposits from the air. In its initial tests, flying over areas that had already been mapped by conventional geologic techniques, *l'avion renifleur*, or "the sniffer plane," was spectacularly successful in picking out oil fields. Realizing that such an invention could alter the course of history, French president Valery Giscard d'Estaing ordered tight government secrecy to maintain France's lead in this new technology.

Over the next three years France invested some $200 million in the idea, but by 1979 government officials were growing nervous. In spite of its demonstrated ability to spot known oil fields, *l'avion renifleur* had yet to uncover any new petroleum reserves. And so far, no government official had even had a close look at the device, having been warned of dangerous levels of radiation. Eventually, the government appointed a prominent nuclear physicist, Jules Horowitz, to investigate.

It didn't take long for Professor Horowitz to devise a simple demonstration. Could these mysterious particles be used to image a metal object through an opaque screen? Yes, of course, he was assured; all that would be necessary would be to "tune" the instrument to the object before it was placed behind the screen. Horowitz chose a simple metal ruler to be the object. He placed the ruler in front of the screen for the tune-up. He then moved it behind the screen, but as he did so, he surreptitiously bent the ruler into an L shape. The device produced a splendid image—but it was of an unbent ruler. Count de Villegas and his associates promptly disappeared. When the device on board *l'avion renifluer* was dismantled, it turned out to be nothing more than a clever video recorder that stored the images of existing geologic surveys.

Tight government secrecy had permitted the deception to go unchallenged for three years. Now that *l'avion renifluer* was exposed as a fake, the original justification for secrecy no longer existed. Ironically, the lid was screwed down even more tightly. The French government now relied on secrecy to avoid embarrassment. It was no longer mere economic dominance at stake; it had become a matter of political survival. Politicians can survive sex scandals or fiscal mismanagement, but they cannot survive being laughed at.

In May of 1981, however, the conservative Giscard was defeated in a runoff election by his liberal nemesis, François Mitterand. It was another two years before the Mitterand government stumbled on the cover-up. Mitterand immediately revoked the secrecy order, gleefully revealing the episode in its entirety, thus dooming any plans of Giscard's to again seek the presidency.

What we cannot know, of course, is how many similar episodes in all countries have never been exposed. Only the censor knows for certain what is hidden. As the story of Project Mogul demonstrated, secret programs can escape public exposure for decades.

TEN

HOW STRANGE
IS THE UNIVERSE?
In Which Ancient Superstitions Reappear as Pseudoscience

NEWTON WAS WEIRD

AGELESS BODY, TIMELESS MIND: The Quantum Alternative to Growing Old, by Deepak Chopra, M.D., was at the top of the *New York Times* best-seller list week after week in 1993, even while *Quantum Healing,* also by Chopra, was still in the top ten from two years earlier. The promise of both books was that illness and even the aging process can be banished by the power of the mind. If anyone should doubt it, Dr. Chopra explains that it's all firmly grounded in quantum theory. Commenting on the occasional spontaneous remission of cancer, for example, Chopra explains: "Such patients apparently jump to a new level of consciousness that prohibits the existence of cancer . . . this is a *quantum jump* from one level of functioning to a higher level." Lest you imagine he was using *quantum* in some metaphorical sense, he informs the reader: "Once known only to physicists, a quantum is the indivisible unit in which waves may be emitted or absorbed, according to the eminent physicist Stephen Hawking."

Physicists wince at Chopra's use of the word *quantum* in the context of a discussion of cancer. That he would cite as an authority Stephen Hawking, who was not yet born when the concept of a quantum originated, suggests that Dr. Chopra's familiarity with quantum theory consists of having read Hawking's enormously popular book on cosmology, *A Brief History of Time*.

Nevertheless, Deepak Chopra's message appealed to millions of intelligent, educated people who have come to believe that we live in a universe so strange that anything is possible. Why wouldn't they? As if a universe of quark matter and black holes were not strange enough, they can read about parallel universes, worm holes through spacetime, quantum teleportation, and ten-dimensional superstrings. Speculative ideas are an important part of the scientific process. Although many of these ideas will not survive, failing either to account for what is already known or to predict as yet undiscovered phenomena, intellectual exploration stretches the imagination even when it fails. But the distinction between untested (or even untestable) speculation and genuine scientific progress is often lost in media coverage.

The confusion is reinforced by the scientists themselves. They are eager to tell people what it's like on the frontier. They want to talk about neutrino oscillations, Higgs bosons, cosmic inflation, and quantum weirdness—the things that excite them. And of course they should—this is part of the human adventure—but in doing so they cannot resist pandering to the public's appetite for the "spooky" part of science. It often seems that the underlying message the nonscientist takes away is that the universe defies common sense, that anything is possible.

Why does the universe revealed by science seem so strange? Our intuitive sense of what the universe should be like is determined by the scale on which we live our lives: the length of a step and the brief span of a human life. Our scientific instruments, however, allow us to study the universe on very different scales. On those scales, "the world is not only stranger than we imagine, it is stranger than we can imagine," as the British geneticist J. B. S. Haldane commented. This famous quote is often used to support the notion of a mysterious universe beyond our understanding. That, I think, is not what Haldane was trying to say at all. He was making a far more profound point: we find it impossible to imagine that which we cannot experience.

Science cannot claim to offer a literal description of underlying reality. We must rely on images drawn from our senses on the human scale to describe things on the scale of the very small, or the very massive, or the very fast. Otherwise we have no way to think about these things. Trouble enters when we insist that all of nature conform to the rules of our human experience. Quantum mechanics, for example, treats matter and energy both as particles and as waves. The idea that matter and energy could be described both as particles and as waves seemed at first to be so strange as to be ridiculous—and it may be. But waves and particles serve our purpose by providing a mathematical structure that allows us to predict how nature will behave.

As we become accustomed to most ideas, of course, they gradually lose their strangeness. What could have been stranger, after all, than Newton's theory of gravity? It involved the mysterious notion of "action at a distance," in which gravity acted instantaneously across the entire universe. That was difficult to reconcile with the principle of cause and effect. But by the time Albert Einstein's general theory of relativity resolved the paradox in 1916, by replacing action at a distance with the curvature of spacetime, the world had spent 230 years getting accustomed to Newtonian gravity. It was Einstein's theory that now seemed outlandish.

The idea that gravitational attraction results from the deformation of spacetime, however, is not really such an unfamiliar concept. Leaves floating on a still pond in the forest gradually collect together in "rafts." They attract each other with a force that results from a phenomenon much like the deformation of spacetime. The weight of a leaf, supported by surface tension, produces a slight depression in the smooth surface of the pond. Two leaves on the surface of the pond will be drawn to one another like two people in a bed that sags. But there is no instantaneous "action at a distance." When a new leaf lands on the pond, the tiny disturbance it creates spreads across the surface as a ripple. The presence of the new leaf has no effect on other leaves on the pond until that ripple reaches them.

In Einstein's general theory, spacetime is deformed by the presence of a massive object, much as the surface of the pond is deformed by a floating leaf. A ripple in the fabric of spacetime, called a gravity wave, can travel only at the speed of light. Action at a distance is banished; causality is restored. The planets in their motions are merely sliding along on the frictionless contours of space. What is it that is curved? How can space, which is empty, have contours? The theory doesn't answer such questions. But it allows us to predict, within well-defined limits, just where those planets will be as far into the future as we choose to look. That's the aim of science: to make the universe less strange, but only in the sense that it becomes more predictable. And in that sense, the universe is not nearly as strange as it used to be. The message the public should take away is that it is not the psychics and fortune-tellers who can see into the future, it is the scientists.

MARS ASCENDING

In the fall of 1997, ABC's *Good Morning America* program carried a three-part series: "Fringe or Frontier? Science on the Edge." The correspondent was Michael Guillen, the Ph.D. physicist turned TV science editor, whom we met earlier reporting on the Patterson cell. On successive days, the "Fringe or Frontier?" series dealt with: precognition, or the belief that humans have premonitions of events that are about to occur; astrology, or the belief that the positions of celestial bodies influence human affairs; and psychokinesis, or the belief that inanimate objects can be controlled by human thoughts.

Two hundred years ago these ideas were associated with witchcraft and sorcery. Educated men and women at the time thought the most important contribution of science would be to free the world from this kind of superstitious nonsense. It was not to be. At the close of a century of incredible scientific progress, the fraction of the population holding these magical beliefs may actually be growing. And yet, if you ask people if they are superstitious, most will deny it indignantly. They have been persuaded that there is real scientific evidence for these ideas. Clothed in the language and symbols of science, these superstitions have survived in a scientific world by mimicking the very agent that was expected to eradicate them. Superstition becomes pseudoscience. Since both magic and science produce remarkable effects that may seem inexplicable, how can the nonscientist be expected to distinguish one from the other?

They would get no help from *Good Morning America*. The first segment of "Fringe or Frontier?" looked into precognition and telepathy. It began with a parade of unidentified talking heads relating personal anecdotes of premonitions that seemed to be borne out. Psychologists point out that there are two things going on here. In the first place people quickly forget the times when they had such feelings and nothing happened; a noncoincidence does not activate the belief engine. In the second place, people respond to subtle clues; the wonderful pattern-recognition equipment of the brain recognizes connections that we may not be consciously aware of. Reading body language is a good example. Most people become very good at distinguishing false smiles from genuine but are hard put to explain just what it is they look for.

Scientists place great value on "scientific intuition," a sort of instinctive sense of how nature will behave under some set of conditions. Some scientists seem to have more of this than others, but they all realize that "scientific intuition" is not some innate sixth sense. It results from experience with natural phenomena that may have followed a similar pattern—much as a skilled billiards player "knows" how to strike the ball without carrying out any calculations of colliding bodies. "Intuition," to the extent that it exists, is simply pattern recognition.

Parapsychologist Dean Radin at the University of Nevada, however, believes there is something more going on. On *Good Morning America* he tested Michael Guillen's reaction to a series of images on a computer screen; some images were calm and soothing, others very disturbing. By measuring the amount of sweat on Guillen's skin, Radin claimed, he could judge the level of Guillen's emotional arousal. Radin contends that emotional response to a disturbing image often starts before the image appears, as if the sub-

jects are able to anticipate what is coming—even though the subjects may not themselves be aware that they are disturbed.

He makes no claim that these premonitions are 100 percent accurate. Intuition, it seems, gets it right more often than it gets it
wrong, but only by a slight statistical margin. You will recall from
chapter 2 that one of Irving Langmuir's symptoms of pseudoscience was that the evidence often consists of a narrow statistical
margin, with no way of widening that margin.

The design of the experiment raises a host of questions. Is simply
measuring skin moisture a valid test of emotional arousal? And why
all this elaborate business of the computer images anyway? Why
not just ask the subjects to "see" what card will come up next in a
deck? The reason, one suspects, is that the experiment of reading
cards can be easily replicated by others. Indeed, it has been repeated
many times and under carefully controlled conditions shows nothing beyond chance. It is much more difficult to refute the computer
image experiment. If another researcher fails to confirm such an
effect, you can imagine endless arguments over what sort of images
must be used, the proper interval between images, the influence of
the subject's background, etc.

It is just one more variation in a tiresome history of extrasensory perception studies. In 1987, at the request of the U.S.
Army, the National Academy of Sciences undertook a complete
review of all the literature on parapsychology as part of a larger
study of unconventional methods of enhancing human performance. The report concluded that there was "no scientific justification from research conducted over a period of 130 years for the
existence of parapsychological phenomena." That has not
changed, except to add more years of failure. As one set of experiments is shown to be flawed, however, new, more bizarre experiments, like the one demonstrated on *Good Morning America,* are
devised. Hope springs eternal. There is always the promise that
the next study will finally produce convincing proof of extrasensory perception.

The more important question for the present discussion is
whether the study of precognition is science. Suppose we apply the
definition of *science* from chapter 2:

Science is the systematic enterprise of gathering knowledge about
the world and organizing and condensing that knowledge into testable laws and theories.

Is there a theory of precognition that can be tested? Guillen appeared to think so: "As to the theory, it's as if we see time through
a very narrow slit. According to conventional science we see only
one moment in time, the present. But, says Radin, maybe the slit
is wide enough to let in not only the present, but a little bit of the
future, causing us now and again to have gut feelings about the
future."

What could he be talking about? We experience the world
through our senses. We understand those senses today in exquisite
detail. To "see" is to record photons as they strike the retina of the
eye, producing a photochemical reaction. The optic nerves transmit
this information to the visual cortex of the brain for processing.
Because photons must travel to the eye, from the thing observed,
we actually see only into the past. Indeed, light reaching our eyes
from distant galaxies may have started its trip billions of years ago.
It is literally meaningless to talk about "seeing" into the future.

But is there some sixth sense feeding information to the brain
that science has not identified? If so, it uses some detection system
very different from the mechanical devices that make up the other
senses. It is also notoriously unreliable. Why do people have acci-
dents if we have some sixth sense warning us?

For the segment on psychokinesis, Michael Guillen visited the
Princeton laboratory of Robert Jahn, a retired dean of engineering
from Princeton. Jahn was, in fact, the mentor of Dean Radin. For
eighteen years, Jahn has been conducting experiments on psychokinesis in which people attempt to influence the behavior of simple
"random" machines. "It's a very small effect," Jahn explains, "not
large enough that you're going to notice it over a brief experiment.
But over very long periods of study, we see a systematic departure
of the behavior of the machine in correlation with what the op-
erator wants it to do." By now you have come to expect this—a
large number of trials with a tiny statistical deviation from pure
chance, and apparently no way to increase the strength of the effect.

In this variation, Guillen is seated at a table with a toy mechanical frog on it. The frog wanders about the table randomly while Guillen attempts by mental effort to get it to move toward him. Sometimes it does—sometimes it doesn't. When it does, Guillen feels he's controlling the frog. Similar feelings lead thousands of people to disaster at the craps tables of Las Vegas. When the frog goes the wrong way, Guillen feels he has lost concentration. When it's over, Jahn announces that computer analysis of the data shows a 90 percent probability that Guillen exerted some influence over the frog.

Why, you may wonder, all this business of random machines? Jahn has studied random number generators, water fountains in which the subject tries to urge drops to greater heights, all sorts of machines. But it is not clear that any of these machines are truly random. Indeed, it is generally believed that there are no truly random machines. It may be, therefore, that the lack of randomness only begins to show up after many trials. Besides, if the mind can influence inanimate objects, why not simply measure the static force the mind can exert? Modern ultramicrobalances can routinely measure a force of much less than a billionth of an ounce. Why not just use your psychokinetic powers to deflect a microbalance? It's sensitive, simple, even quantitative, with no need for any dubious statistical analysis.

The reason, of course, is that the microbalance stubbornly refuses to budge. That may explain why statistical studies are so popular in parapsychological research: they introduce all sorts of opportunities for uncertainty and error. And error has a way of seeming to support the biases of the experimenter. In the random-walking frog experiment, for example, the outcome might be influenced by the clock that starts and stops the experiment. If the clock is consistently started when the frog is moving toward the subject, it will provide all the statistical margin that is needed to declare a positive finding. The researcher might not even be conscious of bias in starting the clock. It should be pointed out, however, that the experiments are often carried out by assistants, who quickly learn that positive results put the boss in a good mood.

This brings up another symptom of pathological science to be added to those listed by Langmuir: there does not appear to be anything resembling progress. The evidence never gets any stronger. Decades pass, and there is never a clear photograph of a flying saucer or the Loch Ness monster. Ten years after the announcement of cold fusion, results are no more persuasive than those obtained in the first weeks. No proof of psychic phenomena is ever found. In spite of all the tests devised by parapsychologists like Jahn and Radin, and huge amounts of data collected over a period of many years, the results are no more convincing today than when they began their experiments. No mechanism is ever uncovered. No testable theory ever emerges.

In each segment of "Fringe or Frontier?" there was, of course, the token scientific skeptic, explaining in a fifteen-second sound bite that such ideas have no basis in science. As Langmuir discovered to his dismay in debunking the ESP experiments of J. B. Rhine more than sixty years earlier, confronting pseudoscience has a way of seeming to take a dispute between superstition and science and elevate it to a simple disagreement between scientists. The more famous the challengers, the more stature they seem to lend to the pseudoscience.

The token skeptic in each segment was always followed by a credulous Michael Guillen. "You have to take it seriously," he gushed after Jahn's experiments on psychokinesis. "If he's right, you can almost foresee a future of mind-controlled wheelchairs, computers, and jet fighters. This stuff is really interesting."

In the segment dealing with astrology, Guillen interviewed a German parapsychologist named Suitbert Ertl, who dismissed most astrology as nonsense but then claimed to have verified that something known as the Gauquelin effect, also called the Mars effect, is valid. Michel Gauquelin, a French psychologist, claimed in 1955 that people, or at least Parisians, born when Mars is rising or directly overhead are more likely to become athletes, military officers, or executives; in the case of Saturn, scientists and physicians are more likely; with Jupiter, it's actors.

Curiously, however, according to Ertl, the Gauquelin effect only holds for those who achieve celebrity status. That seems quite undemocratic, but it's consistent with the observation that Hollywood celebrities who make contact with their past lives invariably find that they were celebrities then as well; they were Cleopatra or Napoleon, never a cobbler or a beggar. It's rather as though the universe is designed solely for the famous, and the rest of us exist only to pay them homage.

Astrology, whether it's the Mars effect or some other form of predestination linked to the stars, does not by any stretch resemble our definition of science. And yet, in summing up the astrology segment of "Fringe or Frontier?" Michael Guillen offered the following: "I don't believe in astrology, and this effect is so small that it affects, if it is true at all, about one in ten million people. But on the other hand, very small effects in physics and astronomy— quantum mechanics, the theory of special relativity—have borne out. So I think we're just going to have to suspend judgment."

Suspend judgment! About astrology? Uniquely positioned to help millions of scientifically unsophisticated viewers understand how the natural world behaves, Guillen chose instead to portray the darkest superstitions that beset our species as open scientific questions—and he did so by invoking the strangeness of modern physics. Let's look at some of the strange ideas of modern physics and see if they offer any scientific support for these ancient superstitions.

THE CASE AGAINST BUTTERFLIES

Could the planet Mars have some effect, albeit small, on life on planet Earth? If the answer is to be found in the laws of physics, the only possible candidate is gravity. In principle, the gravitational field of a celestial body extends throughout the universe. The gravitational pull of the Moon certainly affects life on Earth through the tides. It has even been argued that life may have originated in tidal pools, where the compounds necessary for life would be concentrated.

If lunar gravity affects the tides, could it have some effect on our bodies or minds? Do human beings experience tidal forces? To answer that question, we must first understand the origin of tidal forces. The tides result from the *difference* in the Moon's gravitational pull on opposite sides of the Earth. That's why there are two tides each day. Tides in lakes are too small to observe because the distance from the Moon to the surface of the lake is so little different than the distance to the lake bottom. Humans, therefore, are much too short to be affected in any way by tidal forces. Indeed, if you stand under an apple tree, the tidal force from an apple above your head, because it is so close, would be much greater than the tidal force the moon exerts on you.

Newton deduced that the tidal force depends on the *cube* of the distance between the centers of two bodies. That is, if the Moon were twice as far away from Earth, it would produce a tidal force eight times smaller. Mars, however, even at closest approach is about 140 times more distant than the Moon. So although Mars is eight times as massive as the Moon, the tidal force from Mars is about three hundred thousand times smaller than that of the Moon. This raises an important question: is a force ever so small it can be ignored? The answer is an unequivocal yes. It has to do with temperature.

A stream or pond that is quite clear in the winter months may become turbid when the weather warms up. Tiny particles of foreign matter that settled to the bottom under the force of Earth's gravity when the water was cold remain suspended in warmer water. If a drop of the water is examined under a microscope, the suspended particles seem to be executing a dance. This is known as Brownian motion, and its mathematical description was one of Albert Einstein's early scientific contributions. It is due to the constant buffeting of the suspended particles by water molecules. This is thermal energy, and in the case of suspended particles, it overcomes even the effect of Earth's gravity. The thermal energy of any substance is dictated by the laws of thermodynamics and is simply proportional to the temperature. At the temperatures of the Earth's surface, the feeble tidal force of Mars is a noneffect, wiped out by thermal energy.

Failure to take thermal energy into account is one of the most frequent mistakes leading to pathological science. It was the failure to account for thermal energy, you will recall, that led some scientists to believe that power-line fields could induce cancer. Nowhere is the failure to take thermal energy into account more evident than in the explanations given for the supposed "memory of water" that is crucial to the claims of homeopathy. It is somewhat puzzling that homeopathists place such importance on water's memory in the first place, given that most over-the-counter homeopathic medications, as we saw in chapter 3, are sold in the form of tablets or even chewing gum, not aqueous solutions. However, ignoring for the moment the eventual fate of the water, what is the basis for the claim that it can retain some memory of substances once dissolved in it? The memory is usually said to reside in the "structure" of the water. There has been an abundance of speculation about what sort of "structure" this might be: clusters of water molecules arranged in specific patterns, or even arrangements of isotopes such as deuterium or oxygen-18.

Such ideas seem to be based on "snapshots" of the molecular arrangement in a liquid. Water molecules attempt to arrange themselves in an orderly fashion, and that is just what happens if the temperature is reduced to the freezing point: ice crystals form in which molecules take up fixed positions. But above the freezing point, the weak bonds between molecules are broken apart by molecular motion—the same thermal energy that produces Brownian motion. If you could take a photograph of the water molecules at some instant, regions would be seen resembling small ice crystals. But these regions of order are transient, forming only to be broken up an instant later by the vibrations of the molecules. Tiny ordered regions are constantly forming and disintegrating. That not even local order can persist beyond the briefest of "relaxation" periods is the definition of a liquid. Hardly the stuff of memory.

Homeopathists respond that very little memory is required. In his book *Healing with Homeopathy*, for example, Wayne Jonas, the director of the NIH Center of Alternative and Complementary Medicine, invokes chaos theory as a possible explanation for water's memory:

> One concept in chaos theory is that very small changes in a variable may cause a system to jump to a very different pattern of activity, such as a small shift in wind direction drastically affecting climatic patterns of temperature and precipitation. Under this way of thinking, the homeopathic remedy can be seen as a small variable that alters the symptom pattern of an illness.

Jonas appears to have borrowed his weather metaphor from the famous question posed in 1960 by meteorologist Edward Lorenz, "Does the flap of a butterfly's wings in Brazil set off a tornado in

Texas?" Working with computerized simulations of weather, Lorenz had found that extremely tiny differences in his initial conditions, such as temperature distribution and wind currents, resulted in wildly different outcomes as the simulated weather evolved over time—outcomes that could not be predicted in advance. His rhetorical question, however, was misunderstood by Jonas. Lorenz was not suggesting that such small effects can be used to control the weather.

On the contrary, by carrying out an enormous number of simulations, Lorenz found that although the results never repeated themselves, they tended to cluster around certain values called attractors, which can be widely separated. You can imagine a drop of rain falling on the Continental Divide. The slightest perturbation can influence whether that drop winds up in the Atlantic Ocean or the Pacific. The complex conditions that determine which set of outcomes a chaotic system ends up in are, by their nature, unpredictable. Not only does chaos theory fail to provide support for homeopathy, chaos offers proof that homeopathy cannot possibly work, any more than tornadoes can be prevented by eradicating butterflies.

HEISENBERG WAS CERTAIN

Quantum mechanics, it has been said, is a test. If someone says he understands it, it means he hasn't thought about it deeply enough. Quantum phenomena seem to require that observations made at one place affect what will be observed someplace else at the same instant—thus appearing to violate Einsteinian causality. Attempts to reconcile this puzzling "nonlocal" behavior with the characteristics of the world we perceive with our senses have engaged the great minds of physics, including Einstein and Bohr, in a grand debate that has never been resolved. For a time, the debate seemed to be put on hold; most scientists were just too busy using quantum mechanics to worry about why it works. Measured by the incredible range of phenomena it renders comprehensible and the technologies it has spawned, quantum mechanics must surely be the most successful scientific theory in history.

But the unresolved questions still gnaw at many physicists. The great physicist John Wheeler, now almost ninety, pleads for a renewal of "the desperate sense of puzzlement" that characterized those early years of quantum theory. Such a renewal may now be underway as physicists struggle to make use of quantum nonlocality in the emerging field of quantum computing.

Much of the confusion about quantum mechanics in the mind of the public comes from a misunderstanding of the Heisenberg uncertainty principle, which is often taken to be a statement that the world is unpredictable. Just the opposite is true. The uncertainty principle is actually a recipe for making measurements with a precision that would be unimaginable classically.

Suppose, for example, that you want to clock the speed of an automobile. You could set up two pylons on the side of the road a known distance apart. An observer at the first pylon will press a button that starts a clock running when the car passes; when the car passes pylon number two, a second observer will press a button stopping the clock. Dividing the distance between pylons by the time on the clock, you now have a measure of the speed of the automobile. The accuracy of the measurement depends on such things as how precisely the pylons are positioned and how quickly the observers respond. The effect of these uncertainties can be minimized by simply using a larger separation between the pylons.

But now suppose you also ask where the car was when its speed was measured. The answer is "between the pylons." The more accurately you determine the automobile's speed by moving the pylons apart, the less precise you can be about its position. If you want to be more precise about the position of the car, you must move the pylons closer together, making the speed measurement more uncertain.

This trade-off is the classic dilemma of measurement. Position and motion are said to be "complementary" variables. There are all sorts of complementary variables in our lives; if we look for greater security in investments, for example, we must settle for a lower rate of return. That was understood before the quantum revolution, but it was supposed that you could always improve the measurement with better instruments.

What Heisenberg postulated was that there is a fundamental limit on how accurately you can simultaneously know both the position and motion of a particle. That limit, called the Planck constant, is a measure of the ultimate graininess of nature. The result, however, is to limit the possible outcomes of an experiment. A quantum transition between two states of an atom results in the emission of a photon of very specific energy. The same transition will always result in a photon of precisely that energy. Heisenberg had actually made the world more certain.

Quantum theory represents the properties of a particle by a mathematical expression called the wave function, which is used to calculate the probability that a particle will be found at a particular position. In keeping with Heisenberg's uncertainty principle, a particle with a well-defined state of motion is represented by a very broad wave packet. Once the particle is detected, the wave function is said to have "collapsed" to the location of the detector. The act of observing a particle has thus caused the wave function to change everywhere. It is as if, until it was detected, the particle was everywhere at once.

Most physicists shrug their shoulders and ask, "Who cares?" Quantum mechanics gives them a mathematical description of nature that accounts for the outcome of their experiments. But for all the power of the two great scientific revelations of twentieth-century physics, general relativity and quantum mechanics, it has not been possible to reconcile them with each other. General relativity is a classical continuum theory, which conceives of the universe as a seamless whole. Therein must lie the source of its incompatibility with quantum mechanics. So far, in spite of many attempts, there is no generally accepted quantum theory of gravity.

When Einstein published his general theory of relativity in 1916, its predictions on a laboratory scale differed so little from Newtonian mechanics that some physicists despaired of laboratory confirmation; it seemed impossible to make measurements with sufficient precision. It is a wonderful irony that confirmation became possible, almost routine, through technologies ushered in by the quantum revolution—the atomic clock, for example, which is accurate to one second in a hundred thousand years. The atomic clock is regulated by the frequency of microwaves emitted during quantum transitions of cesium atoms. So much for the shibboleth that quantum mechanics describes an unpredictable world. Because its predictions are faithfully borne out by experiments, quantum mechanics has made the world of the scientist much less strange.

That mysteries remain is no surprise. That quantum mechanics has not been reconciled with general relativity only reminds us that there are great discoveries still to be made. When that happens the world will become even less strange.

THE UNCONSCIOUS UNIVERSE

Not everyone is happy with a predictable world. In such a world, humans seem to be reduced to nothing more than complex machines, ultimately governed by the same laws of physics that keep the planets in their orbits. The worldview of most people requires some sort of human essence that transcends our physical bodies—a soul, perhaps—that makes humans more than just machinery.

Some seek evidence of this special human essence in science and claim to find it in quantum mechanics. They believe quantum mechanics describes a world that responds to human consciousness. For them the wave function becomes more than a mathematical construction; it is given a physical reality—a sort of holistic consciousness field that permeates the entire universe, transcending time and space. By the act of observing the output of an instrument, the human observer has affected the wave function of some system, a change that reaches instantly across the universe.

This concept is sometimes given the grand title of "participatory anthropic principle." Suppose that there is no human observer, and instead the output of the instrument is recorded. According to the participatory anthropic principle, the wave function would not be collapsed until a human observer consciously examines the recording. Until that happens, the event registered by the instrument would only be a "potential" event. Nothing can truly be said to have happened until it is observed by a human. Our minds literally create reality.

It's rather hard to prove this is not so, but it does not seem to me to be a very useful way to look at the world, and it raises all sorts of awkward questions: How conscious must the human observer be to resolve potentiality into actuality? Can all humans collapse wave functions? If a chimpanzee is trained to make the observation, could the chimpanzee collapse a wave function? What about a human who is no smarter than a chimpanzee?

While there are sober scientists who accept some version of a participatory anthropic principle, it is enthusiastically embraced by those who seek evidence of paranormal phenomena. In Deepak Chopra's *Ageless Body, Timeless Mind: The Quantum Alternative to Growing Old* we read that "the physical world, including our bodies, is a response of the observer . . . beliefs, thoughts and emotions create the chemical reactions that uphold life in every cell." This is a tenet of ayurveda, the traditional religious healing of India, which goes back thousands of years. Quantum theory is invoked by Chopra to convey the impression that ayurvedic medicine has somehow been validated by modern science. We cannot help but notice, however, that the author of *Ageless Body* shows unmistakable signs of growing old right along with the rest of us.

In *Healing with Homeopathy*, Jonas takes this concept a giant step further than even Chopra:

> Some theorists suggest that intentionality and consciousness must be brought into any explanation of how nonlocal quantum potentials might be "collapsed" into molecules . . . Thoughts or beliefs nudge potential effects into existence by an intention to heal in the person or practitioner.

He is proposing that molecules needed for healing can be created by the thoughts not only of the patient but of the healer. While it is of course true that our emotions can, within limits, influence the chemistry of our bodies, to grant such power to the thoughts of the healer is a fantastic leap. (I refrain from calling it a quantum leap.) It conjures images not only of psychic healing but of casting spells. Where once the magician in his robes would have called forth the spirits, the pseudoscientist invokes quantum mechanics, relativity, and chaos.

THE YOUNGEST SCIENTIST

The power of the healer is even more explicitly summoned in "biofield therapeutics," commonly known as touch therapy, though it would be more accurate to call it no-touch therapy, since the practitioner's hands do not actually make contact with the patient. Instead, the practitioner's hands are moved over the body an inch or two away, smoothing and balancing the patient's "energy field." The nature of this presumed energy field is not very clear, although it is sometimes said to be electromagnetic. Practitioners agree that the field extends outward from the body for several inches. The claim is that a trained touch therapist can tactilely sense this energy field.

Touch therapy has become very popular in recent years and is now available for patients who request it in more than seventy hospitals across the nation. Adapted from the ancient Chinese practice of *qi gong*, touch therapy was introduced into the United States by Delores Krieger, a professor of nursing at New York University. It is often offered as an adjunct to surgery, and some surgeons report it calms and relaxes patients. Certainly, there is no evidence that anyone has been physically harmed by having his or her aura manipulated.

There is no evidence either of therapeutic benefit, other than testimonials from satisfied recipients. Touch therapists explain that double-blind trials are not possible. It seems that the benefit results from the biofield of the practitioner coming into confluence with the recipient's biofield. The ability to perform biofield healing is therefore universal, although most people seem unaware of possessing the talent. As with any innate talent, results improve with training and practice, but because the biofield is always there, it is not possible to perform sham healing. The recipient always benefits.

Emily Rosa, however, decided it would be possible to do a double-blind test of the claim that touch therapists can feel the body's energy field, which is variously described as a tingling sensation, or warmth, or a gentle resistance. She persuaded twenty-one local touch therapists in Boulder, Colorado, to submit to a beautifully simple double-blind test. How could they refuse? Emily was a charming nine-year-old schoolgirl who wanted to conduct the test for a fourth-grade science fair project.

Emily spent ten dollars on materials for the test. She would be seated across a table from the subject. The table was divided by a screen so they could not see each other. The therapist would extend both hands, palms down, through holes in the screen. Emily would then place one of her hands just below, but not touching, one of the therapist's hands. A coin toss would be used to decide under which of the therapist's hands she would place hers. The therapist was supposed to identify the hand that was in the presence of Emily's energy field.

Once the procedure was explained, many of the therapists expressed confidence that they would be able to sense the presence of Emily's hand with 100 percent accuracy. The entire procedure was captured on videotape. In 280 trials, the therapists scored 44 percent. The therapists were stunned. They were honestly convinced of their ability to sense a human energy field, but in a double-blind test the power had deserted them.

With the help of her mother and a medical statistician, Emily's experiment was written up and submitted to the prestigious *Journal of the American Medical Association*. After a thorough review by expert statisticians, the editors of *JAMA* declared the study to be "solid gold." Emily became the youngest scientist ever to publish a paper in a major medical journal. She was showered with honors and appeared on nationwide television news. The James Randi Educational Foundation gave Emily a grant of a thousand dollars toward her next research project—one hundred times as much as she spent on the touch therapy investigation.

I recently met Emily's mother at a medical conference in Philadelphia. She told me that Emily, who had just turned twelve, was now devising a test of magnet therapy. Emily Rosa, the scientist, is doing what scientists are supposed to do—taking the strangeness out of the universe.

THE COSMIC ZOOM

For a million years, our species was confronted with a world we could not hope to understand. Now, almost within the span of a single human lifetime, the book of nature has been opened wide. On its pages we are finding, if not a simple world, at least an orderly world in which everything from the birth of stars to falling in love is governed by the same natural laws. Those laws cannot be circumvented by any amount of piety or cleverness, but they can be understood. Uncovering them should be the highest goal of a civilized society. Not, as we have seen, because scientists have any claim to greater intellect or virtue, but because the scientific method transcends the flaws of individual scientists. Science is the only way we have of separating the truth from ideology, or fraud, or mere foolishness.

Ideology, fraud, and foolishness were all present in the examples of voodoo science discussed in these pages. For those who intentionally set out to commit fraud, such as the makers of "Vitamin O," we can have little sympathy. But most of the scientists and inventors we met started out like Joe Newman, believing that they had made a great discovery overlooked by everyone else. While it never pays to underestimate the human capacity for self-deception, they must at some point begin to realize that things are not behaving as they had supposed.

Like all those who have gone down this road before them, they will have reached a fork. In one direction lies the admission that they may have been mistaken. The more publicly and forcefully they have pressed their claim, the more difficult it will be to take that road. In the other direction is denial. Experiments may be repeated over and over in an attempt to make it make it come out "right," or elaborate explanations will be concocted as to why contrary evidence cannot be trusted. Endless reasons may be found to postpone critical experiments that might settle the issue. The further scientists travel down that road, the less likely it becomes that they will ever turn back. Every appearance on nationwide television, every new investor, every bit of celebrity and wealth that comes their way makes turning back less likely. This is the road to fraud.

Few if any scientists are so clever or so lucky that they will not come to such a fork in their career. Almost all will acknowledge their error and put it behind them. Some will start down the road of denial but recognize in time that they are headed in the wrong direction and turn back. A surprising number—apparently unable to face turning back, and yet unwilling to follow the road all the way to fraud—seem to leave the road entirely, completely losing touch with reality. No matter how difficult it becomes to keep believing, it is easier than facing the truth.

They pose no great threat to science. Voodoo science is a sort of background noise, annoying but rarely rising to a level that seriously interferes with genuine scientific discourse. Something like cold fusion might interrupt the flow of science for a few months, but those who make extraordinary claims must eventually produce the evidence. The more serious threat is to the public, which is not often in a position to judge which claims are real and which are voodoo. Those who are fortunate enough to have chosen science as a career have an obligation to inform the public about voodoo science.

Most people who are drawn to voodoo science simply long for a world in which things are some other way than the way they are. Some cannot accept that we are prisoners of the Sun. They look wistfully at the stars that fill the night and imagine that there *must* be some way to overcome the limitations of space and time. Others refuse to believe that the dreams and emotions that stir within them can be reduced to the laws of physics. They seek in science some evidence of a cosmos that cares about *them*. All that scientists can do is to explain what we have learned, and we have learned a lot.

In 1996 I attended a showing of *Cosmic Voyage*, a new IMAX film at the Air and Space Museum in Washington that captured the state of our knowledge of the universe. I was unprepared for the film's emotional impact. Would it, I wondered, have a similar effect on nonscientists? When I returned to the office, I asked my secretary if she would view the film and give me her impressions. Delia is well read and sensitive to the human condition, but she has no background in science. She agreed, not knowing what to expect.

She was almost overcome. As the "cosmic zoom" hurled viewers to the outer limits of the universe, plunged them down to the domain of the quark, and sent them tumbling back through billions of years, one factor of ten at a time, she was terrified—terrified by her growing realization of the insignificance of Earth and its creatures. Galaxies collided, stars exploded, worlds were obliterated. Humans were powerless before such forces. But terror mingled with wonder. Wonder that fragile, self-replicating specks of matter, trapped on a tiny planet for a few dozen orbits about an undistinguished star among countless other stars in one of billions of galaxies, have managed to figure all this out. That is perhaps the strangest thing about the universe. Strange and very wonderful.

INDEX